The Llano Estacado of the
US Southern High Plains

UNU Studies on Critical Environmental Regions
Edited by Jeanne X. Kasperson, Roger E. Kasperson, and B.L. Turner II

Note from the editors

This book is the sixth in a series from the United Nations University (UNU) research project, Critical Zones in Global Environmental Change, itself part of the UNU programme on the Human and Policy Dimensions of Global Change. Both endeavours explore the complex linkages between human activities and the environment.

The project views the human causes of and responses to major changes in biochemical systems – global environmental change broadly defined – as consequences of cumulative and synergistic actions (or inactions) of individuals, groups, and states, occurring in their local and regional settings. The study examines and compares nine regional cases in which large-scale, human-induced environmental changes portend to threaten the sustainability of an existing system. The aim is to define common lessons about regional trajectories and dynamics of change as well as the types of human actions that breed environmental criticality and endangerment, thereby contributing to global environmental change. The overall results of the comparative analysis are found in *Regions at Risk*, the initial volume in this series.

Titles currently available:

- Regions at Risk: Comparisons of Threatened Environments
- In Place of the Forest: Environmental and Socio-economic Transformation in Borneo and the Eastern Malay Peninsula
- Amazonia: Resiliency and Dynamism of the Land and its People
- The Basin of Mexico: Critical Environmental Issues and Sustainability
- The Ordos Plateau of China: An Endangered Environment
- The Llano Estacado of the US Southern High Plains: Environmental Transformation and the Prospect for Sustainability

The Llano Estacado of the US Southern High Plains: Environmental transformation and the prospect for sustainability

Elizabeth Brooks and Jacque Emel
with Brad Jokisch and Paul Robbins

United Nations
University Press

TOKYO • NEW YORK • PARIS

The views expressed in this publication are those of the authors and do not necessarily reflect the views of the United Nations University.

United Nations University Press
The United Nations University, 53-70, Jingumae 5-chome, Shibuya-ku, Tokyo, 150-8925, Japan
Tel: +81-3-3499-2811 Fax: +81-3-3406-7345
E-mail: sales@hq.unu.edu
http://www.unu.edu

United Nations University Office in North America
2 United Nations Plaza, Room DC2-1462-70, New York, NY 10017, USA
Tel: +1-212-963-6387 Fax: +1-212-371-9454
E-mail: unuona@igc.apc.org

United Nations University Press is the publishing division of the United Nations University.

Cover design by Joyce C. Weston

Printed in the United States of America

UNUP-1042
ISBN 92-808-1042-1

Library of Congress Cataloging-in-Publication Data

Brooks, Elizabeth.
The Llano Estacado of the US Southern High Plains: environmental transformation and the prospect for sustainability / Elizabeth Brooks and Jacque Emel, with Brad Jokisch and Paul Robbins.
p. cm.
Includes bibliographical references and index.
ISBN 92-808-1042-1
1. Llano Estacado-Economic conditions. 2. Llano Estacado-Environmental conditions. 3. Sustainable development-Llano Estacado. 4. Irrigation-Economic aspects-Llano Estacado. I. Emel, Jacque. II. Jokisch, Brad. III. Robbins, Paul. IV. Title.
HC107.T42 L62 2000
333.7′09764′8–dc21 00-010504

Contents

List of figures and tables

Figures

Tables

Acknowledgements

We would like to thank the staff of High Plains Underground Water Conservation District #1, the librarians and staff of the High Plains Collection of the Texas Tech University Library, and Otis Templer of Texas Tech University for all their invaluable help and time. We would also like to thank the many government officials, agency staff, and private citizens we consulted or interviewed in the Lubbock area for their insights and information. Lastly, we would like to thank the editors of this series and the anonymous reviewers for their comments.

Preface

The "region at risk" in this volume is a portion of the U.S. Southern High Plains, an area straddling the Texas-New Mexico border and bounded on the north by the Canadian River in the Texas Panhandle, on the east and south by the Caprock Escarpment, and on the west by the Pecos River Valley (Brooks and Emel, 1995: 256). The arid, featureless, and windswept landscape prompted early Spanish explorers to brand the region El Llano Estacado – "the Staked Plain" – signalling their perception of a forbidding environment. According to Mexican, Tejano, and ultimately Anglo-Texan lore, the Llano was the last refuge of the uncivilized. It attracted the feared *Comancheros*, marauding outlaws and renegades who plagued "civilized society" hundreds of kilometres to the west, along the upper Río Grande of New Mexico, and to the south-east in the Edwards Plateau of Texas.

The reputation of the Llano was not solely a figment of European culture. The southern end of the High Plains is part of the so-called "Texas cultural sink," a term some anthropologists and archeologists have variously enlisted to denote the vast area between the highly developed Mesoamerican cultures of central Mexico, the settled Pueblo of the upper Río Grande, and the Mississippian cultures who penetrated the waterways into the eastern Great Plains. To be sure, the "sink" is open to challenge on various grounds: its very designation constitutes interpretive biases; it and the Llano were occupied throughout their "prehistory"; and increasing evidence indicates that Amerindian farmers once settled along some of the rivers cutting through the Llano. Yet the height of Amerindian civilization on the Llano awaited the "horse culture" that arose in the seventeenth century, after Spanish introduction of the domesticated horse into the Río Grande basin to the west. Given new mobility, various Indian groups, none more so than the Comanches, followed the great bison herds into their southern and winter range in the Llano.

The Llano was, for the most part, avoided by Spaniards and Mexicans during their political dominions over what is now the southwestern United States. Mexican occupation as late as the 1800s focused on southern Texas (Tejas) and the upper Río Grande from El Paso to Taos (now in Texas and New Mexico, respectively). "Colonization" of Texas by Anglos from the United States and later by various groups directly from Europe increasingly pushed occupation north and west towards the Llano. By the mid nineteenth century, a string of forts to the south of the Llano demarcated the frontier of "settled" Texas, whereas forts along the southern extension of the Rocky Mountains set the perimeter for Río Grande settlements.

These conditions changed radically in the late 1860s with the end of the American Civil War and the rush of Anglos to "develop" the west. The penetration of the central Great Plains by railways opened markets for beef from the millions of longhorns that roamed the landscapes of northern Mexico and southern Texas, a product of a long-established Spanish and Mexican industry and land use. Thus began in earnest the era of great trail drives northward to the railheads in Kansas and the establishment of the Anglo ranching culture of Texas and the south-west. The Llano's short but open grasslands became home to some of the largest ranches in America.

Farmers soon followed, but rain-fed cultivation was risky in this semi-arid region. The turning point came in the 1930s with the introduction of turbine pumps and the mining of the groundwater of the southern Ogallala aquifer. A contested issue of the current day, irrigation marked the modern transformation of the Llano into a rich farming/cattle-raising region. Also, the rich deposits of oil found on the southern and western edges of the Llano in the 1920s added a powerful industrial component to the regional economy.

As in much of the American West, the transformation of the

human-environment condition of the Llano was as swift as it was radical. In less than a century, the horse culture of the Plains Indians succumbed to Anglo ranchers and farmers and, ultimately, to urban-industrial America. A grassland seasonally roamed by bison herds surrendered to fenced property under livestock and cotton. Everywhere the road and rail networks, tying rural land uses and users to the rapidly developing cities of Amarillo and Lubbock, penetrated the Llano, facilitating the industrialization of farming and ranching, complete with large-scale irrigation, hybrid crops, fertilizer, and pesticides. These and information networks linked the land users to international markets as well as to "expert" knowledge from high-powered "Ag-schools" and state and federal "Ag-agents." Western society, in short, transfigured the once forbidding Llano into a region of wealth and political power – a transformation symbolized by the proliferation of glowing light towers illuminating the Friday-night ritual of high-school football across the vast expanse of the region.

If this were the end of the story, the Cornucopian and technological-fix perspectives would be vindicated. Increasing technology and socio-economic change, even in the face of rapidly growing demands for natural resources, changed a land marginal for human use to a bountiful one. The story of the Llano would fall into the "transformation success" category mentioned by George Perkins Marsh in 1864 (Marsh, 1965) and could be called upon to demonstrate the power of market demand to increase productivity and resources. Ecocentric and preservationist concerns might be relegated largely to the loss of the Llano's biotic diversity, especially its fauna – a concern which garnered little sympathy on a Llano dominated by the ethic of "humankind over nature."

But, of course, the story of the Llano is a continuing one, and signals abound to trigger the warnings by Cassandras and conservationists that the Llano is poised to bite back, or at least to cease to provide. Strip away the sophisticated technologies that support the scale of its occupation and use, and the Llano remains a sem-iarid to arid region affected dramatically by decadal and longer climatic flux. The relatively ephemeral and nomadic use of the Llano by Amerindians constituted one adaptation to the vagaries of nature and to chronic water scarcity. The Anglo farmer registered another, albeit poorly adapted, mechanism. Entering the Llano during a period of higher than usual rainfall, farmers soon suffered the reverberations of inappropriate land management in the face of extended and extensive drought. The 1930s were devastating to Llano farmers as they watched their parched topsoil, carried aloft by the strong winds of the Plains, become the substance of the Dust Bowl. Many smallholders were devastated. Emigrating in all directions, they were as much the inspiration for Steinbeck's *Grapes of Wrath* as were his "Okies" (Oklahomans).

The adaptation to this socio-economic and environmental trauma, shared with others throughout the Dust Bowl period, was improved land management and increased use of technology, as farmers and ranchers worked closely with agricultural scientists and extension agents. No amount of manipulation could deliver an economically acceptable solution, however, without improved supplies of water, and that water lay underground for the taking. The Llano's hidden treasure was, and remains, the Southern Great Plains or Ogallala Aquifer, a storehouse of fossil water some 50–100 metres thick collected from post-Cretaceous times by the eastward movement of Rocky Mountain groundwater. The advent of turbine pumps provided plentiful and inexpensive supplies of irrigation water. The Dust Bowl Llano became an oasis for cotton production and linked much of the regional economy to international markets for this fibre. Institutions controlling access to water, at least in the Texas portion of the Llano, fostered rapid and almost unregulated pumping of the Ogallala. Profits soared, and with them the economic and political clout of Llano inhabitants, whose influence in state and federal policies steadily grew.

Economic growth and the rise of urban-industrial centres increased water demands. The water-mining far exceeded the recharge rate of the aquifer, however; and the water level of the Ogallala began to drop, precipitously in many cases. Deeper and deeper pumping increased the costs of water, with impacts constituting a new "drought" in agriculture. Perceiving this dilemma in the 1960s, some power brokers on the Texas portion of the Llano sought to have the state and national governments supply the region's water by way of far-distant waterworks. One proposal would have invested millions of dollars to pump water from the lower Mississippi River, more than 1,200 kilometres to the south-east (as the crow flies)! It proved too difficult, however, to convince the broader public to underwrite such costly schemes. Alternative strategies were clearly in order.

Today, the rural economy of the Llano stands at a crossroads of these strategies. Agricultural adaptations that use more efficient but expensive irrigation are under way. These costs, the vicissitudes of the international beef and cotton markets, and the economic restructuring of American agriculture gave rise to the family corporate farm. Overall, however, the economic backbone of the region is shifting to its urban-industrial centres. The Llano is experiencing yet another transformation, one which may decrease agricultural land and promote less water-demanding cultivation but which will also increase urban-industrial water demand. Much Llano land may thus be "left alone," perhaps reverting to the short grasslands it once was. The demands on the Ogallala, however, present another issue.

This story and more are detailed by Elizabeth Brooks and Jacque Emel in part of the Project on Critical Environmental Zones (ProCEZ), which produced an interim volume, *Regions at risk: Comparisons of threatened environments* (Kasperson, Kasperson, and Turner 1995). The present volume extends and amplifies one of the chapters (Brooks and Emel, 1995) in *Regions at Risk* and is the sixth volume to appear in the series UNU Studies on Critical Environmental Regions. Tracing the human–environment history of the region, Brooks, Emel, and colleagues pay special attention to the transformations that developed from the late nineteenth century onwards with the entry of Anglo cultures and economies: the loss of the Llano's biodiversity, especially fauna, with the advent of fences, livestock, and plough; the degradation of the land and the decline of smallholder farmers owing to the Dust Bowl droughts; and the depletion of groundwater attendant on the economic resurrection of the region through irrigated cultivation and urban-industrial growth.

This tracing, however, is not restricted to the average or general conditions. Brooks and Emel delve into the socio-economic inequities of the changing human–environment conditions, inequities intimately tied to government policy, resource institutions, and international markets. Access to and control of water are pivotal, and the disempowered are disadvantaged. The smallholder farmer is on the decline relative to largeholders and corporate landholders, who possess greater access to the capital and political influence that allow rural land uses to be economical in the short run. The ebb and flow of the international prices for commodities, especially wheat, cotton, and meat, puts farmers at risk and hampers governmental efforts to maintain various policies that subsidize and support land uses. Increasingly, the growth in urban areas, including their industry and services, challenges rural land uses for water.

Thus Brooks and Emel offer a narrative open to many interpretations: the Llano as a region dramatically changed in the face of increasing human well-being in general; a region environmentally degraded, in which those individuals and groups with economic and political clout increase their wealth; or a region teetering on the brink of a slippery slope and heading down a path of dependency on external political, economic, and technological forces. Their conclusions are that the Llano constitutes a "partially threatened region," one in which, on average, increased material well-being has been and continues to be achieved for a growing population through an export-oriented economy. This achievement, however, has come at the cost of a regional environmental drawdown, particularly in terms of the aquifer, and it is not clear that the current uses are sustainable over the long run.

The authors' narrative illustrates the adaptations to this drawdown and to climatic drought, almost all of which involve technological solutions (e.g. pumping) or economic adjustments (shift to urban-industrial sectors) that have increased regional production and consumption. The growth of a significant urban population and industrialization, however, stresses water demands and places the agrarian sector in competition with the urban-industrial sector for this critical resource. Farmers have responded by moving to more water-efficient but more costly modes of irrigation. How long this adaptation will prove sustainable is unclear. It is plausible that the technologies of water retrieval will become so costly as to require major adjustments in the regional economy, with unforeseen implications for regional land and resource use and for levels of occupation and affluence.

In this sense, the Llano Estacado is "partially threatened" in the overall scheme of the comparative study reported on in this series. The aquifer has been and remains the critical resource. The impacts of its drawdown within the greater political economy of the region could lead to several human–environment futures. One extreme vision foresees a Llano returned to its historical, marginalized position with reduced economic well-being measured by modern standards of human settlement occupancy and production. No Comancheros exist in this vision, but the Friday-night glow from the athletics fields may dim considerably. An alternative extreme envisions significant restructuring of the regional economy based on less water-demanding activities, perhaps focused on "high-tech" industries supported by state universities. In this vision, stress on landscapes is reduced while human well-being is sustained, even improved. Many variants exist between these two visions.

Interestingly, the Llano is situated in a political economy – technologically advanced, wealthy, capitalist-controlled with government interventions – that could readily lead to futures tending toward either extreme. The vagaries of the market and the changing location of capital could lead to a new, marginalized Llano, a mined landscape abandoned. These same attributes, coupled with wise policy, could lead to more environmentally benign use of the Llano, and even a return of some of its landscapes to native biota, while increasing regional well-being economically. Which of these visions will typify the Llano's future depends not only on the "hidden hand" of political economy but on the political directions espoused by regional interests.

Jeanne X. Kasperson
Roger E. Kasperson
B.L. Turner II
Series Editors

1

The Llano Estacado and the question of sustainability

"Across the Path of Empire lay a great, empty, forbidding land, open to every wind that blows." ... It was a "worthless, inhospitable region," said the explorer Marcy in 1866, "destined in the future, as in the past, to be the abode of wandering savages." (Johnson, 1947: 19)

The Llano Estacado

Vast, nearly featureless, and perceptibly high, the Llano Estacado of Texas, a subregion of the US Southern High Plains, is a daunting place. Its climate is challenging, with incessant winds that can fluctuate from soft springlike breezes to ferocious howling blizzards and back again in the span of a week. The intractable tendency toward drought can be read in the landscape: towering windmill skeletons, abandoned farms and town sites, a roadside drift of fine red sand. The horizon constantly recedes; the sky seems wider and closer. But the Llano is also a place of great beauty, with vast, barely undulating plains, infrequently interrupted by a small stand of trees, or a distant townscape. Lone farmsteads linger on in vaguely sheltered corners of long-ago farm fields.

Large cities rise up as well. Lubbock is a relative metropolis, with a population of just under two hundred thousand. This "Hub of the

Plains" is an important commercial centre for cotton processing and oil and gas production, with regional dominance in higher education and medical services. Amarillo, a smaller city to the north (150,000 population), is perhaps better known, thanks to its early days as the starting point of nineteenth-century cattle drives to the packing houses of Kansas City and Chicago. Its reputation as a cowboy town has been slowly replaced by its more recent incarnations as a centre for cattle finishing, high-technology weapons manufacture and decommissioning, and agriculture, and as the site of the US Strategic Helium Reserves. There are many smaller towns in the region: the Llano is a curious mix of wide-open agricultural lands and urban centres. Small towns such as Earth and Morton have gradually disappeared over time on the Llano but without significant population loss overall (Bauer, 2000). People have migrated to the bigger cities, where, the hope is, opportunities still exist as the possibilities of the agricultural economy shrink.

Although the region has been populated for more than 10,000 years, the settled, modern Llano is a recent phenomenon. Lubbock, the regional capital, was founded in 1907. The first permanent Anglo-European settlement started in 1879. Within a generation, a profound environmental transformation began. Land so suitable to the production of grasses seemed naturally suited to the production of domesticated grasses and fibre plants as well. The end of the 1800s and the early years of the new century were an era of technological innovation and institutional optimism. The confluence of those broader trends with the "opening" of the Llano set in motion the wide-scale agriculturalization of the region. Mechanized tractors, bulky and expensive, were ideally suited to flat, treeless, and rock-free plains. And the years of good weather that marked the turn of the twentieth century ensured a steady flow of capital onto the Llano, making investment in the expensive new technology possible.

The prosperity and expansion of the early part of the twentieth century was punctuated by the swift descent into the "Dirty Thirties," an era of crushing economic depression and devastating environmental repercussions of the overuse and overextension of land, people, and capital. But just as the bad times came, they left, and with the industrial rebirth that followed the end of the Second World War, the Llano rebounded with wild expansion into the new frontier of irrigated agriculture, dependent upon a seemingly inexhaustible source of groundwater, the High Plains Aquifer System (locally known as the Ogallala Aquifer). Within a relatively short time period

(100 years), then, an enormous shift in the ecology of the region occurred and nearly all of the land cover was transformed. From relatively sparse nomadic use of the region by the Comanches, the region has been altered to facilitate commodified systems of crop and animal production, to the point of the introduction of bio-engineered and genetically altered crops. This massive ecological shift engineered by human ingenuity – reflected in technology development, institution building, economic power, capital investment, and educational achievement – has allowed for the growth and maintenance of a modern industrial agricultural system that has earned significant wealth for the region. Yet one of the critical elements for this system, the ancient and irreplaceable groundwater, is non-renewable and nearly exhausted. As a result, the Llano is again facing change, neither major nor sweeping as before, but incremental and erosional, as water levels decline and pumping costs increase. The economic life of the water supply is forecasted variously as between 10 and 50 years; those forecasts probably have built in variability of at least that many years. In the parlance of the larger Project on Critical Environmental Zones (ProCEZ), of which this study is a part, the region holds to a shifting line between an "impoverished" and an "endangered" zone. The difference between the two corresponds with the timeline for reaching the point at which "continuation of current human-use systems or levels of well-being" are precluded and the possibilities for different future uses are narrowed (Kasperson et al., 1995: 25). Endangerment refers to situations in which the non-sustainability of human-use systems may occur within the next generation, whereas impoverishment refers to a longer timeframe (beyond the next generation) for reaching a non-sustainable level.

The purpose of this book is to examine the forces that brought the Llano, and the people making their homes and livelihoods there, to this situation of proximate endangerment. Following the work of other environmental historians, we frame the study within a political-economy approach which explains the transformation of the region as a result of the development and practices of major institutions: markets, governments, property, and elements of civil society (see for example Worster, 1979, 1985; Cronon, 1983, 1992). We are informed by the studies of political ecologists as well. After Blaikie and Brookfield (1987), we are concerned with the pressure of production on resources and the opportunities and constraints of the land manager. We also consider the roles played by culture, ideology, and human inspiration and inventiveness in guiding or creating the environmental and social history of the region. With their focus on marginalized peoples seeking to maintain or attain livelihood from a region that is being integrated into the global economy, political ecologists, had they existed in the 1930s, might have studied this area because of the general poverty and ecological degradation of the drought years, culminating in the Dust Bowl; poverty, or at least the struggle away from poverty, certainly was a central fact in the ecological degradation of that era. We diverge from the political ecology approach with our focus upon the historical transformation of a region that willingly embraced integration into global markets from a position of considerable poverty and duress.

One might conclude that today's problems with groundwater depletion are the result of the successful decisions and institutional practices that ameliorated the problems of poverty and soil erosion. People of the region are now faced with a problem of relative wealth and "overexploitation" of the resource base. The efforts to "fix" the problem of the 1930s have generated new problems of sustainability for this region. "Development" has already occurred. Could more sustainable ways of creating livelihoods in the region have been imagined?

Many studies of sustainability (or its inverse, "criticality") have focused upon third world communities and ecosystems (Peet and Watts, 1996). To think of sustainability in the context of a relatively long-industrialized, globally embedded agricultural region is not necessarily more challenging, but requires a somewhat different set of considerations. The wholesale transformation of the ecology that predated European settlement has already taken place. A highly capitalized and technologized agricultural system with attendant urban centres has already been created on the basis of an exhaustible water resource. The problem is not what can be preserved of the pre-European or "pre-modern" ecology, but what can be maintained of extant communities and livelihoods – what roads may be taken to ensure the sustainability of as many future livelihoods as possible, for those who wish to stay in the region. Short of bringing in water from outside the region, a prospect that is currently unlikely because of the interregional tensions regarding water transfers and big dams, the only obvious solution is to discontinue or greatly reduce irrigated agriculture and industrialized beef finishing. Depletion of the water by continued high levels of agricultural withdrawal guarantees diminishment of the larger cities' potential for economic diversification. But preserving the water supply for future municipal use may be undermined by loss of the agricultural foundation of the region's economy.

What will the region look like a hundred years from now? Tomorrow may bring further transformation to an unknowable ecological or social state. Will the wealth generated by water disappear with the exhaustion of the resource, or will other modes of economic existence grow to replace the intensive water-based agricultural economy? How will agricultural communities fare in this shift from one type of economic production to another? Who will gain and who will lose? Who will stay and who will go? And, finally, what conclusions can be drawn about this environmental and economic transformation of the past 300 years? How might it be evaluated or judged and by what criteria? To begin, we turn to the idea of the "region" itself and not only lay out a physical description of the Llano Estacado but articulate our assumptions regarding the ways in which "regions" are both physically and socially constructed and understood. Then we turn to the concept of "sustainability" to map out the multiple meanings of the term so that we may refer to them throughout our examination of the environmental and economic transformation of the region. We close this chapter with an overview of the organization of the book.

The definition of regions

The Llano, "broad, slightly rolling to undulating lands," has existed as a physiographically defined region since the ancient runoff of the then-forming Rocky Mountain system of North America "deposited the smooth coalescing alluvial-fan plains of vast extent" (Johnson, 1931) that created these austere highlands (see figure 1.1). As with all regions, however, the southern plains exist as much more than a physical entity. The peoples who have inhabited and traversed the plains for thousands of years have each defined the region for themselves, and in turn have been defined by the region they created. The region is a physical entity; it is also a social construct, created both by the inhabitants of the plains and by the rest of the nation.

The concept of regionality

A casual neglect of the careful consideration of what exactly defines a region has been notable in the practice of regional analysis. Some regional scholars have raised the question of how regions are derived, differentiating among the classic terms of region as subdivided space

Figure 1.1 **The Llano Estacado**

with some functional purpose, space, locality, or territory. Whereas conventional regional study has been characterized by its acontextuality, functionalism, and ahistoric explanation (Murphy, 1991), regions are most completely understood as constructs, socially defined entities existing in a physically defined space. Regions become "spatial constructs with deep ideological significance that may or may not correspond to political or formal constructs," produced through an iterative process of history and current context (Murphy, 1991). The study of regions should also be undertaken as the study of the people living within a defined region, with careful explication of their physical and historical context. The endless negotiation among the individual members of the group, the negotiation of the group with the context, and the constraints on that negotiation from the sociocultural and political structures within which these components exist describe the complexity of regionality (Thrift, 1991: 461).

Moreover, the question of how deeply social structures are influenced by the specifics of context needs to be addressed in consideration of the meaning of a region. It has been noted that general processes which are recognizable despite being particularized by their specific expressions in a given regional setting may in fact be so thoroughly redefined by the context that they are, in the end, not so general (Sayer, 1989). In other words, land distribution laws might appear to be modified for the semi-arid plains lands, but closer analysis can reveal those laws to be the medium for perpetuating the same general principles (individual ownership of privately held property, yeomen farmers) that attended their inception and made them impracticable for any region, regardless of climate or land use. Recognizing how context-dependent any given phenomenon may or may not be is essential to determining those potential characteristics of regionality that actually have meaning for describing the region.

Beyond the problematic nature of the definition of spatial, cultural, and physical regional boundaries is the issue of interpretation of experience once it has been locationally fixed. One group's understanding of its history in a place will almost certainly differ from another's; the implications of this for the definition of a place are considerable. Research that ignores this complexity risks failure: "behaviour seems intelligible only to the extent that it resonates, not because we have articulated the constitutive structure of understanding or feeling" (Sayer, 1989: 256). The explanation of a region's history and present conditions must strike a balance between narrative and analysis; the reconstruction of the past in a narrative prefigures the depiction of current situation and conditions. And yet, without analysis of the abstracted trends and forces that are widely recognized, what is left is a mere retelling of events. Analysis without narrative, however, results in narrow description without a sense of competing forces and interpretations.

The socially defined region

Regions are both physically and socially constructed. Donald Worster, an eminent historian of the plains, defines a region as "the outcome of a dialogue between culture and nature" (1994, xi); the Southern High Plains is surely such a construction of a particular set of social facts confronting certain immutable aspects of the environment. And the adaptations of law and custom by people set in an ideological context, confronting a set of environmental circumstances, are explored below. Moreover, the West, as the larger contextual region, can be read as a "battleground between the global economic system of capitalism, which was amenable neither to environmental adaptation, conservation, nor democracy, and an alternative social ideal of public planning, communal ownership of resources, and community decision making about their development" (Worster, 1994: 25). About the Southern High Plains specifically it has been said: "the Plains have been an endless puzzle and a considerable disappointment to those who have tried to tame them" (Worster, 1994: 91) – or possibly, to define them.

Describing areas of semi-arid to arid climate, with few trees and little surface water, dominated by cotton and cattle, as a "social unit ... [with] measurable characteristics, largely determined by the dominant structural and functional aspects of the region," Cleland (1966: 2) uses the term "a sociological region," as opposed to a strictly physiographic region. He goes on to question whether the fact of aridity (or semi-aridity) has further implications for the definition of regions. Such regions must be comprehended, he concludes, through the multiplicity of sociocultural variables brought to prominence by the fact of aridity (or semi-aridity). Adaptations to water-deficit areas, including those measures that seek to negate the impact of insufficient water, have varied over time and economic space. People living on the Southern High Plains today structure their daily lives quite differently from their predecessors of 150 years ago or even 50 years ago, and yet they are still constrained by the same physical facts of the region.

Arguably, ways of life in communities in largely agricultural regions can be distinctly different in semi-arid and humid regions regarding practices such as "land use, water law,... political attitudes,... and other aspects of the culture including folklore, attitudes, and values of the people" (Cleland, 1966: 3). Goldschmidt's (1978) intensive study of two irrigated agricultural communities in the Central Valley of California, one consisting primarily of family farms, the other of agro-industrial farms, illustrates how economic forms can produce different societal forms within the same semi-arid area.

Comanche history (which predates Euro-American history in the region) records a different physical world from that described by Euro-American settlers; the same physical setting was bounded very differently for the two populations as well. Environmental phenomena, for example a period of drought, would have significantly different resonance with an agrarian community than with a nomadic herding group. Thus, an essential aspect of regionality is who lived in a given space at a given time. The competing histories of the region are the result of the several groups of humans who inhabited the plains over a period of about 10,000 years, beginning with paleo-hunter and gatherer groups and ending with the group currently populating the plains, the descendants of immigrants and other Euro-Americans who moved onto the plains around the turn of the twentieth century. Each of these groups defined the region differently, and had different relationships to the environment in which they made their homes. The earliest inhabitants defined the region quite locally: a particular lakebed, or riverine environment. Later, with the coalescing of nomadic groups, the region expanded to cover the area traversed in a year's cycle. The introduction of the horse brought an entirely new conception of the range of the region inhabited by the Native Americans. And, lastly, the agriculture-based colonization of the region fixed forever the spatial limits of the Southern High Plains.

The overlapping human experiences on the Southern High Plains, and in particular that area known as the Llano Estacado, have contributed substantially to the region's claim to uniqueness. To become one of the last regions "settled" by European Americans, the Southern High Plains were essentially depopulated and stripped of the physical characteristics which had defined them. The region was then made over to the demands of the newest inhabitants. This was a pattern that was repeated throughout the American landscape (for example, see Cronon, 1983).

With the preceding in mind, the physical characteristics of the Southern High Plains become the starting point for both narrative and analysis. Taken together with the unfolding of events in the social history of the area in the following chapters, this analysis allows the understanding of the region as a totality of location and experience.

Physical characteristics of the Llano Estacado

Physiography

The Southern High Plains, bordering the eastern face of the Rocky Mountains of North America, overlay an immense network of subterranean water deposits known commonly as the Ogallala Aquifer.

We focus on a particular segment of the High Plains, a vast region spanning the Texas Panhandle–Plains known as the Llano Estacado, or "Staked Plains," so named after a possibly apocryphal tale of Spanish explorers so bewildered by the featureless undulating plains that they placed stakes in the ground to mark their route and thus a way back to their starting point. Specifically, this region is bounded on the east and south by the Caprock Escarpment, a series of bluffs and cliffs cut back by several small surface streams, on the west by the Pecos River valley in New Mexico (but, for this study, bounded on the west, albeit artificially, by the state line), and on the north by the Canadian River, north of Amarillo in the Texas Panhandle. The Llano Estacado was formed by alluvial deposits from ancient rivers "which spread the sediments in broad fan-like forms in some cases hundreds of feet in thickness" and rises gently from about 600 metres (about 1,970 feet) above sea level on the easternmost boundary to about 1,200 metres (about 3,950 feet) on the westernmost edge (Johnson, 1931). The average rise is slightly more than three metres (10 feet) of elevation per 1.6 kilometres (1 mile) (Fenneman, 1931). The area comprising the Llano measures about 90,650 square kilometres (35,000 square miles), about 11 per cent of the land area of the state of Texas (Urban, 1992).

Climate, soils, water

The Llano Estacado is semi-arid, with low rainfall and a long growing season, characterized by one of the highest percentages of sunny days in the continental United States and a long frost-free period. Rainfall amounts can be rather low, ranging from about 510 millimetres (mm) (about 20 inches) annually on the eastern reaches of the Llano to about 360 mm (about 14 inches) annually along the western edge of the region. May and September are generally the periods of peak rainfall, with very dry conditions existing from October through April – an important period of time for the replenishment of stored soil moisture. The traditional boundary of subhumid and semi-arid climates in North America, the 510 mm isohyet (20 inches) annually (Johnson, 1894), falls slightly east of the centre of the Llano, designating the region as one of relatively deficient rainfall. During the major periods of recurrent drought in this century, from 1930 to 1960, the 510 mm isohyet moved significantly to the east across the Llano (Texas Agricultural Experiment Station, 1968). In the 1990s, conditions on the Llano illustrated again the variability of the climate as the region underwent a drought considered by some to be the worst of the twentieth century, with rainfall amounts under 250 mm (less than 10 inches) for a period of nearly 12 months in 1996–1997.

Moderate temperatures and a fairly late occurrence of killing frosts (typically early November) contributed to the perception that the Llano was a natural agricultural region. Relatively early springs, defined by last frost generally occurring in mid-April, and soil temperatures reaching 10 degrees Celsius (50 degrees Fahrenheit) by the end of March, helped shape the region's agricultural base. All told, the Llano offers a growing season of about 185 days in its western reaches and 225 days to the east. The region also receives a very high percentage of possible sunshine – one of the highest in the contiguous United States, about 70 per cent annually (Potts, Lewis, and Dories, 1966).

An important agricultural climatic factor for the Llano, however, is the virtually continuous wind. The seasonally very windy conditions have desiccating effects on young crops (the windiest period of the year tends to be from March to May, when planting occurs) and exacerbate soil erosion, a serious issue throughout the region. The prevailing wind direction is west and south-west for half the year (late autumn through mid-spring) and south for the rest of the year (mid-spring through late autumn) (Lee et al., 1994). Studies such as those carried out by Lee and colleagues (1994), however, have concluded that wind patterns are largely chaotic and highly variable as to direction and speed.

This is a region of rich, dark brown to reddish brown soils, mostly sandy loams, clay loams, and some areas of clayey loams of uniform texture (Livingston, 1952). These are soils created from the Quaternary aeolian sand and loess sheet known as the Blackwell Draw Formation (Lee et al., 1994). All fine-textured, with the exception of the coarser sandier soils, these soils are very susceptible to wind and water erosion, particularly after the removal of the native vegetative cover (Tharp, 1952). These are also soils with a very high organic matter content, which Odum characterizes as those that support the "granaries of the world," assuming adequate rainfall (Odum, 1971: 130). On the Llano, however, "the very forces which have made it possible for these soils to be so high in the chemical materials that give high fertility have also imposed limitation upon their productivity," specifically "... the low rainfall and occasional droughts" that limit the leaching of organic material (Johnson, 1931: 73–4).

Other critical factors concerning soils on the Llano include their high susceptibility to wind erosion, particularly the sandier soils of the western and southern parts of the region (Tharp, 1952). Additionally, at varying depths, a thick layer of *caliche*, or hardpan, a clayey-limy stratum, appears throughout the Llano, indicative of lake deposit stratigraphy (Darrow, 1958), from which the Caprock gets its name (Johnson, 1931).

One of the most important features of the Southern High Plains is the occurrence of extensive groundwater reserves. The aquifer underlying the region is part of the High Plains Regional Aquifer (also known as the Ogallala Aquifer), a vast underground reservoir covering parts of eight states from Wyoming and South Dakota in the north to Texas and New Mexico in the south, with an area of approximately 450,660 square kilometres (about 174,000 square miles). The estimated drainable capacity of the system is approximately 4.01 trillion cubic metres (3.25 billion acre-feet) (Kromm and White, 1992a: 15).[1] The aquifer was created from alluvial deposits of Quaternary or late Tertiary age, resulting in a formation of mixed sequences of clay, silt, sand, and gravel dropped by streams flowing down the eastern face of the then-forming Rocky Mountains. The Ogallala is underlain by impervious shales of Permian to Cretaceous age, therefore only those parts of the formation with adequate thickness yield enough groundwater to sustain irrigation.

The depth of the layers of silt and sand above the impermeable shales, called locally the Caprock, is known as the saturated thickness and ranges from 0 to 1,600 metres (0–5,250 feet) in the South Plains (Cronin, 1969). The mean saturated thickness over the entire aquifer system is about 60 metres (about 200 feet); the mean saturated thickness for the Texas portion of the aquifer is about 34 metres (112 feet) (Kromm and White, 1992a: 16). Owing to pumping for irrigation, municipal, and domestic uses, this thickness has been reduced in some areas of the study area by more than 30 metres (roughly 100 feet).

Recharge to the aquifer is minimal, averaging less than 25 millimetres (1 inch) each year over the region. Less than 1 per cent of the available runoff percolates through the root zone. As a result, the High Plains Aquifer can be considered non-recharging, and an essentially finite water source. Nevertheless, the High Plains Aquifer System provides the water for 20 per cent of all the irrigated acreage in the United States, some 30 per cent of all the water pumped for irrigation (Kromm and White, 1992b). This groundwater mining, the extraction of a non-renewable resource for consumption, led to predictions in the early 1980s, the last years of the golden age of irrigated agriculture on the plains, of 40-year horizons for the feasibility of groundwater extraction from the aquifer (Walsh, 1980), a tenuous vision of sustainability for the region.

The question of sustainability

Given this overriding constraint, what can it mean to have a sustainable way of life on the Llano? In their essay, "Making sense of sustainability," Gale and Cordray (1994) reviewed nine versions of sustainability, while Pretty and Howes (1993), in their study of agricultural sustainability in Great Britain, identified a bewildering 75 definitions. The concept, critics argue, is in danger of losing its meaning in such a plural conversation. In almost all cases, however, sustainability invokes the imperative to maintain something for the longer term, beyond the current human generation. Specifically, the Brundtland Report of the World Commission on Environment and Development contained this widely accepted, albeit broad, definition of sustainability as "development that meets the goals of the present without compromising the ability of future generations to meet their own needs" (Brundtland, 1987: 3). Defined in other literatures, this broad intergenerational notion ranges from the "deferred consumption" models of environmental economics to ecologists' defence of regeneration of the environment over time (Hendee and Stankey, 1973). Beyond this notion of intergenerational promise, the question becomes "What is to be sustained?" Is it the resource, the community, the ecosystem, or some of each of them? And how can a non-renewable resource dependency be construed in any manner as sustainable?

Sustaining the resource

The most immediate interaction with the environment that creates the greatest impacts on the Llano's environment is the agricultural production system. Sustaining the resource base so as to preserve key economic sectors is one definition of sustainability (Gale and Cordray, 1994). Sustainability is thus measured in the ongoing preservation of a "narrow range of ecosystem products defined as economically valuable for existing markets" or specialized goods for some particular niche of the global economy in which the region has a comparative advantage (Gale and Cordray, 1994: 313).

The rationale for such a form of sustainability follows the arguments of resource economics. While uninterested in the environmental or community value of a system, such an approach is amenable to the reservation of the productive resource over time. The broad measure of this form of sustainability is, simply, highest sustainable yield through protection of the productive resource. In the context of the Llano, this means sustained use of the aquifer over time.

Over the decades of the twentieth century, cotton and cattle have emerged as the dominant production systems on the Llano. Which of them has used the groundwater resource to the greatest productive ends and is likely to continue doing do so over time? Is irrigated agriculture or cattle production more likely to support a longer-term sustainable community?

The dominant market products of the region, cattle and cotton, are both water-demanding. Cotton, though brought to the region as a drought-tolerant crop, has become emblematic of irrigated agro-industry in terms of ecology and institutional support (Brooks and Emel, 1995). Cattle production, though less widely recognized as a regional irrigated product, is also dependent on groundwater resources. Water is essential throughout the many stages of the cattle production and feeding process; however, the largest portion of water use comes from the production of feed grains, specifically sorghum.

Sustaining the community

Sustaining the community is a goal of many of the "schools of sustainability." Defined by Gale and Cordray (1994: 316) as "dependent social systems sustainability," this is an explicitly value-laden concept that asserts the importance or primacy of the existing human social configuration, however defined. Strange (1984: 115) suggested that this is the goal of agriculture, to "nourish a renewable pool of human land stewards who can earn a healthy living by farming well." For Pretty and Howes (1993: 8), this form of sustainability must specifically include "a mechanism to maintain existing farm employment and halt the continual decline in numbers of those directly employed on the land."

In the case of the Llano Estacado, these "stewards" who are "employed" on the land traditionally have been found on the family farm, generally characterized by its independence, owner-operation, commercial diversification, and innovation. Identifying these farms and measuring their fate is more difficult than the above definition

Table 1.1 **Population cycles in the Llano Estacado**

Cycle	Years	% population change*	Mode of transformation
Boom	1880–1890	+??	Settlement and homesteading
Bust	1890–1900	–??	Drought and cattle collapse
Boom	1920–1930	132.79	Dry-farming and mechanization
Stagnation	1930–1940	5.79	Dust Bowl
Boom	1940–1960	33.41	Irrigated cotton and beef
Bust	1960–1992	–18.43	Price collapses and consolidation

* Total population change in the region, excluding the urban counties of Hockley and Lubbock (US Bureau of Census)

initially suggests. The modern, capitalized, family farm of the Llano Estacado is now an agro-industrial firm and a government client (although decreasingly so), whose structure and operations are difficult to bound in the farmhouse (Opie, 1993).

The protection of the rural community need not require the maintenance of "farmers in aspic" (Bowers, 1995: 1242), however. Community sustainability does not have to hold the rural cultural landscape to an abstract and pristine standard; a more practical measure would suffice. A general test of community and operator vitality could incorporate the size and variation of the total population over time as well as the maintenance of the number and size of family holdings. The measure here is the maintenance of population, especially under conditions of economic or environmental shock, and the preservation of independent family farm operations in the region.

The population of the Llano region has been severely affected by each of the successive ecological and economic shocks that have struck the region. As shown in table 1.1, the cycles of growth, stagnation, and decline are driven by changes in the weather, the national and international economy, and the mode of transformation brought to bear on the environmental resources.

The punctuated periods of population decline, especially during the Dust Bowl, would probably have been considerably more dramatic had the state not responded with subsidization and institutionalization of farming and ranching. While good and bad years are common to agricultural practice, these long-wave depopulations and repopulations of the region are more typical of a mining community than an agricultural one. Recent research on these population movements demonstrates that even when the overall population has not declined in recent years, it has redistributed itself in aquifer-exploiting

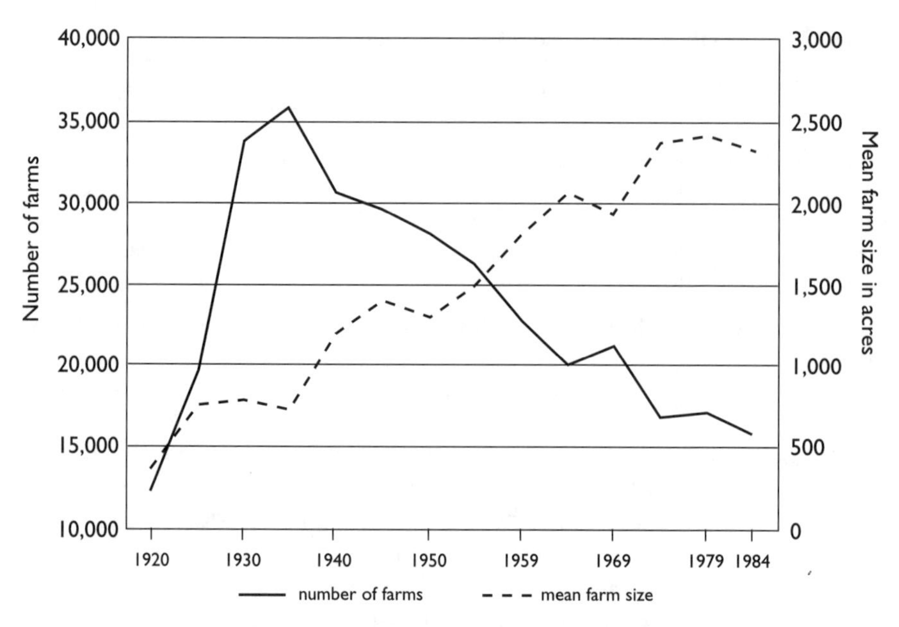

Figure 1.2 **Farm number and size (1920–1984)**
Source: US Agricultural Census, 1920–1985.

pockets; population has declined in some areas while growing in others (White, 1994). The inability of the region to sustain growth stands out in its demographic history, as does the continual transformation of the production system to recapture wealth and draw back producers and consumers. This long-term demographic instability is paralleled by a trend towards consolidation and smallholder decline that also threatens the ability of the region to sustain community.

As shown in figure 1.2, the history of the region is marked by the steady decline in the number of farm operations in the study region (after the early boom years) and the marked increase in farm size. The relationship is a tight funnel, as growth in the size of consolidated farms follows the steady accumulation of land in competitive production. The dwindling population is working in and around an increasingly consolidated agricultural landscape. This trend is consistent throughout the period since World War II, despite subsidies, price supports, and benefits for pumping (Brooks and Emel, 1995). Although irrigation was introduced largely to help the marginal farmer compensate in poor rainfall years, it has become the tool of largeholders in the years following the war.

Sustaining the ecosystem

Sustaining the ecosystem for its own sake is another long-vaunted definition of sustainability. This form of sustainability has been referred to as ecosystem identity or ecosystem benefit (Gale and Cordray, 1994) and it is geared towards the preservation of natural or pristine ecological conditions (Taylor, 1986). Generally rooted in an earth-centred position, it becomes anthropocentric when human interests are viewed broadly and in the long term.

The simplest measurement of this form of sustainability would be the coverage and extent of the original and pristine ecological conditions. Under less stringent conditions, a test of this form of sustainability would measure whether or not a region has been transformed beyond the point of restoration or reclamation.

Although the extent of the original grasslands environment is largely nil, pockets of indigenous nature could conceivably remain, hidden in the interstices of agricultural production. The pre-European range complex that existed in the region (reconstructed by Tharp (1952)), dominated by buffalo grass (*Buchloë dactyloides*), blue grama (*Bouteloua gracilis*), and hairy grama (*Bouteloua hirsuta*), has largely disappeared, but grasslands of some form persist in the open pasturage of the region. The degree to which these pastures have been spared cropping and have retained any of their original biotic qualities is a liberal test of ecological sustainability.

It would be a mistake to characterize these lands as pristine or even as characteristic of grasslands environments. These were, by any standard, destroyed a century ago. These lands are most likely to consist predominantly of long-fallow fields, rotated out of production. The federal government's Conservation Reserve Program, wetlands regulations on playa lakes, and other efforts at rationalizing crop production and conservation have favoured resting of the land. The high percentage of non-cultivated land suggests an ecological reserve of relatively low pressure. Some of the indigenous (predating European settlement) bird populations still persist, as well as smaller mammals and reptiles. If the pre-modern ecology has not been sustained in the region, it has also not been obliterated beyond any hope of recovery, except, of course, for those fully extirpated animal and plant species.

The question of whether the environment has been transformed beyond the point of reclamation is complex. In the case of the semi-arid Llano, this is largely a test of desertification, as broadly understood by range and climate experts (Dregne, 1985). While the irreversibility of this condition is debated, particularly under arid and semi-arid conditions of rainfall variability, it is an important element of true desertification (Grainger, 1990). In the absence of true desertification, it can be argued that the options available for the future are not absolutely limited.

With this in mind, the relatively reduced rate of soil erosion in the region since the 1940s points away from the total and irreversible collapse of the ecology. Irreversible soil erosion has occurred only in a small fraction of the study area, in south-western Gaines and Cochran counties, where hummocky landforms and absolutely barren soils appear (Brooks and Emel, 1995). The total drawdown of the aquifer might conceivably affect a handful of artesian springs but, apart from this, will have relatively little impact on the non-irrigated parts of the surface ecology.

Further, the plant species that dominated in the region – buffalo grass, blue grama, and hairy grama, as well as Wright's three-awn and three-awn grama (*Bouteloua* spp.) – have not been eliminated altogether and do exist as dominant species elsewhere. If land is retired from cultivation, it could be reseeded. For the countless possibly unidentified species that might have been lost in the transformation, re-establishment is obviously impossible. Certainly it would be impossible to recreate pristine conditions, but many of the elements of the grasslands environment persist or could be reintroduced (Popper and Popper, 1994, 1991).

In sum, the agro-economy of the region fails the most stringent tests of ecological sustainability. The wholesale displacement of the indigenous environment was thorough and meticulous. The Great Plains of the Native Americans, as they existed, were destroyed. On the other hand, there is no evidence to suggest that the future range of environmental conditions has been physiographically limited by this century of exploitation. In this sense, sustainable futures might still be imagined, as long as the intractable calculus of a limited water supply is part of those imaginings.

Organization of the book

This book is organized around the major phases of environmental transformation of the Llano. We trace the evolution of the ecological makeup of the region from its **creation** as a vast glacial by-product, through its gradual development as a shortgrass prairie/scrub complex, replete with massive herds of grazing animals, and inhabited by nomadic herders and gatherers. This region existed in slow evolution for thousands of years, populated by large groups of Native Americans who gradually differentiated themselves into tribes with discrete cultures and customs. The arrival of Spanish explorers itself did little to disturb this world, but what this event set in motion profoundly changed the nature of the region.

The second part of this chapter begins with the arrival of the first Anglo traders some 200 hundred years later, accelerating the process of environmental **change**, and the early beginnings of herding and ranching of domesticated animals that set the stage for replacement of native species of animals and plants, as grazing became more extensive.

The first step in the **transformation** of the region, containment of the Native Americans making their homes on the plains, soon became an imperative for the settlement of the region in accordance with Anglo-European desires. This story is told in chapter 3. The US military was a major force in implementing federal-level policies directing the removal and marginalization of the tribes who had lived on the Llano for centuries. The near-extinction of the buffalo, an essential partner in the Comanches' way of life, forced them further to the edges of existence. Within a few decades, agricultural settlers were attempting to set up farming homesteads, replacing the cattle and ranching culture that had been greatly reduced by the cattle bust of the 1880s. Always a cyclical climate veering between severe drought and rainfall sufficiency, the short-term climate began to have a serious impact on the progress of Anglo settlement.

The fourth chapter opens on an era of rainfall sufficiency towards the turn of the twentieth century, an era that ushered in massive environmental change on an unprecedented scale, the equivalent of total **replacement**. The removal of the Native Americans had been an essential aspect of this transformation and in this chapter we explain the replacement of the few remaining indigenous people with European-American agricultural settlers who, within a generation, had embraced large machine technology and succeeded in replacing the indigenous grasslands with grain- and fibre-based agriculture. The upward spiral of development was accelerated by World War I, pushing demand and opening European markets to North American grains and fibre. The replacement ecology proved disastrously nonresilient, and with the onset of post-war drought, and buffeted by the international economic depression, the region collapsed in ecological and financial catastrophe.

The fifth chapter traces the recovery of the region, ecologically and economically, through the discovery of vast supplies of underground water and the availability of technological innovations to access it. This began the era of **exploitation** and **expansion** on the Llano. Agricultural production grew rapidly, with cotton and beef leading in exports and in value. The endless water supply supported the expansion and the great increase in the standard of living for the people on the Llano. The promise the region held for the early waves of settlers was being met.

In the sixth chapter, we assess the current state of the environmental **transformation** of the Llano. Starting in the 1980s, the region has gone through a series of contractions and restructurings, attempting to forge **a new path** into a sustainable future. Prolonged drought, global upheavals, and shifts in international trade have all played significant roles in framing the future of the region and its people.

Finally, we look back over the history of the Llano and the environmental transformation that has unfolded inexorably to the current untenable state. How this region, home to nearly a million inhabitants, can continue **sustaining a way of life** built on an ecological structure of less than a century's making is the issue we have sought to clarify, by weaving together the disparate strands of histories told by succeeding groups of people on the Llano.

Note

1. An acre-foot is the amount of water needed to cover 1 acre (0.4047 hectares) of land to a depth of one foot (about 0.3 metres), or about 326,000 gallons (1,233,910 litres). Alternatively, one acre-foot is equivalent to 1,234 cubic metres.

2

Creation and change: 12,000 B.P. to A.D. 1860

Creation

The Llano Estacado today is the result of millennia of geomorphologic process and a century or so of human endeavour. To fully appreciate the latter, it helps to understand the former. Thousands of years of aggradation are responsible for the rich but vulnerable soils. Historically low rainfall amounts preserved the soils, while supporting the variably scrubby and lush shortgrass terrain. For millions of years, animals and plants evolved to a form of stasis, interdependent and resilient. Over thousands of years, Native Americans gradually made the plains their home, developing nomadic or locally agrarian ways of life so as best to exploit the resources the region offered. Non-extractive resource use made the Native Americans' impact on the Llano transitory, relative to the impacts of later European settlers. Overhunting or overgrazing in one season could be countered by lower reserves of wildlife in the next and a move to another region, giving the indigenous people a chance to survive and the indigenous animals and plants a chance to recover. The symbiosis of the relationship created between the native peoples and their physical environment set the stage for the disruption caused by the incursion of European explorers in the mid sixteenth century.

The Llano Estacado of Texas is a region of stark and forbidding beauty that resulted from hundreds of thousands of years of gradual creation. To the extent that any place ever does, the Llano arrived at a point of natural stasis some thousands of years after the final glaciation, the Wisconsin. Centuries of uplift forming the Rocky Mountains resulted in the deposit of millions of tons of alluvial materials carried away from the eastern face of the developing chain by thousands of streams, rivers, and rivulets.

Plant life

For centuries, prior to Anglo settlement and development, a unique vegetation–wildlife complex existed in the Southern High Plains. Extensive grasslands supported a wide diversity of animals, and provided food for paleo-gatherers living on the plains. Some of the complexity of these ecological communities was uncovered in small pockets of relict prairie communities that were found in the foothills of the eastern Rocky Mountains in the 1950s (Livingston, 1952). However, this vegetative complex no longer exists on the Llano.

During the Late Mesozoic and Cenozoic eras, the great inland seas originally covering the area that eventually became most of the western United States during the Lower Cretaceous period receded. The development of vegetation on the plains is inextricably tied to this stratigraphic history of the region. The "distinctive flora" of the Southern High Plains was derived originally from "tropical or border tropical ancestors" (Darrow, 1958). The aggraded layers of marine sediments remaining when the inland seas receded from what became the Great Plains region supported the predecessor to the modern ecology of the early grasslands complex, which probably emerged during the Oligocene Period, when the south-western climate started to differentiate into drier and less dry regions. This change was concurrent with a general continent-wide cooling trend, which resulted in a drier, non-forested region bordered by forest to the east and west (Darrow, 1958). The actual species that populated the Great Plains region at that time are unknown, but fossil evidence suggests that the region was probably covered by a subhumid grassy scrub vegetation dominated by grasses of the *Stipa* genus; others have posited that instead a tallgrass vegetation complex formed during the Oligocene and included species of the *Andropogon*, *Imperata*, *Eluyonurus*, *Trachypogon*, and *Tripascum* genera (Darrow, 1958).

The grasslands evolved along with major Miocene-era stratigraphic events, chiefly the formation of the Rocky Mountains through block faulting and the general uplift of the Colorado plateau. One result of this massive movement was the eventual deposit of material that formed the High Plains (Darrow, 1958). The ongoing trends of cooling and increasing aridity resulted in further development of grasslands. Some have even suggested that long-term modern-day climatic conditions across the High Plains are analogous to those of the time spanning the Oligocene to the Pliocene (Darrow, 1958). Fossil evidence reveals grazing animals throughout the plains at this point in time, with plants of higher drought tolerance and lower water requirements, largely of the *Andropogon* and *Bouteloua* genera, gradually replacing species that had greater water needs.

Glaciation appears to have had little impact on the further development of the grasslands climax ecology, although the impact on soils is evident through the stratigraphic and fossil record. Darrow (1958) reports that by the end of the Pliocene, clear trends towards increasing aridity showed up in the fossil mollusc deposits, accompanied by decreasing evidence of arboreal flora and even prairie flora. He attributes the Llano's significant *caliche* layer to the changing conditions that signalled increasing climatic difficulty for pre-existing plants and wildlife; since the latest glacial advance, a "distinctive but oscillating trend toward increasing aridity has taken place in the Great Plains area" (Darrow, 1958: 39).

The early modern ecology on the Llano (corresponding to the roughly 10,000 years between the most recent glacial advance and the advent of Hispanic incursion) has been recreated through the use of early explorers' diaries and journals recording what they encountered on the plains, along with studies like Livingston's (1952) examination of relict prairie communities. Although Livingston's relict communities were found on the high plains of Colorado, they are roughly analogous (with some slight variation) to the grassland complexes that probably existed on the Llano. These relict prairie communities reflect "true prairie" complexes and were discovered in pockets interspersed with the more commonly occurring "mixed prairie" climax communities. Livingston found that most relict communities appeared in areas with relatively high water tables. These more mesic environments appeared to be the primary reason the relict communities survived in their scattered locations, possibly even from the time of the glacial retreat. Despite the increasing aridity to the south and west, these relict communities were able to exploit the slightly wetter micro-environments at lower altitudes, in valley bottoms, and in sloughs (Livingston, 1952). Also, some communities appeared to have been pockets of re-introduced species, or to have been maintained by the landowners in their original state for hay production. The relict true prairie communities are characterized by the dominance of certain indicator species, chief among them *Sporobolus heterolepis*, *Stipa spartea*, and *Sporobolus asper* (Livingston, 1952).

By the time of the first recorded ventures onto the Llano, the oscillating climatic conditions referred to by Darrow had trended towards the decidedly more arid. The notion of a Great American Desert in the middle of the North American continent was created by the earliest Anglo explorers, among them Stephen H. Long, who recorded his impressions of his 1819–1821 travels through the vast grasslands for the United States government.

This modern-era vegetation complex was dominated by blue grama (*Bouteloua gracilis*), and hairy grama grasses (*Bouteloua hirsuta*). The secondary species included buffalo grass (*Buchloë dactyloides*). The grama grasses were reduced over time to secondary importance by grazing.[1] One-metre-tall wheat grasses were probably found in micro-environments such as buffalo wallows, sloughs, or sinkholes. Other micro-environments, such as canyon floors and dune areas, accommodated different plant complexes. For example, the canyon floors, cutbacks from the Escarpment, with their less sandy soils and rolling topography, were localized sites of hair-leaved sand sage (*Artemisia filifolia*) with some more common grama grasses such as side oats (*Avena* spp.) and sweet bluestem (*Andropogon* spp.) (Tharp, 1952). The sandier dune areas with sparse grass cover, later stressed by grazing, held more drought-resistant and stabilizing plants such as thread leaf sage (*Artemisia* spp.), covered spike drop-seed (*Sporobolus* sp.) and catclaw (*Mimosa borealis*) (Higgins and Barker, 1982; Odum, 1971; Tharp, 1952). Weed species (sunflowers (*Helianthus* spp.) and tumbleweed (*Salsola* spp.)) increased with increasing grazing and eventually the introduction of ploughing.

The radically restructured current-day vegetative complex of the Llano can be described as largely monocrop agriculture complemented by scrubby, drought-resistant species. Weed species such as sunflowers and tumbleweed have flourished in the niches left vacant by the destruction of the shortgrass species: "conspicuous annual forbs ... such as the Russian thistle tumbleweed (*Salsola*) and sunflowers (*Helianthus*), owe their luxuriance to man's continual disturbance of the soil" (Odum, 1971: 389). Small reserves of grassland are protected under the Forest Service of the United States Department of Agriculture (USDA), through a programme, begun in the 1930s, designed to bring marginal Great Plains farmlands out of agricultural production (West, 1990). Even these preserves, however, were and are treated as resources and are open to grazing.

Animals on the Llano

Climate changes may explain significant alterations in animal life as well. The presence of large lake basins on the Llano Estacado clearly indicates that conditions at some time in the prehistoric past were considerably wetter and cooler than they are at present. Precipitation might have been twice that of the present, with mean annual temperature about 6 degrees Celsius (about 9 degrees Fahrenheit) cooler. After the cooler period came a progressive decrease in effective moisture, probably owing to decreasing precipitation and increasing temperatures. The stratigraphic sequences in a series of sites in draws in the region document this trend, showing a shift from free-flowing streams between approximately 13,000 and 12,000 years before the present (B.P.), to open ponds and lakes between 11,000 and 10,000 B.P., and finally, to an aggrading marsh with little or no open water from 10,000 to 6,300 B.P. (Bamforth, 1988). Evidence of extreme aridity after 6,000 B.P. is found at Blackwater Draw and Rattlesnake Draw in eastern New Mexico and at Marks Beach in Texas, where wells dated 6000–5000 B.P. were dug to a buried aquifer.

Two known important changes occurred with the fauna of the Southern High Plains during the terminal Pleistocene and early Holocene. The first of these was the extinction of most of the large mammal species in the region, which apparently occurred between 11,000 and 10,000 B.P. The reasons for the extinction of the Pleistocene mammals are obscure and sometimes contradictory, but the extinctions are at least in part attributable to the appearance of human hunters. Another major change occurred in the morphology and, presumably, the adaptation of the bison or buffalo, the dominant large ungulate remaining. The modern (post-Pleistocene) Llano was a habitat for a great diversity of wildlife, most of which was permanently displaced by the gradual incursion of Anglo settlers. Some species were made extinct; a very few, under current conditions, are regaining a place in the Llano's ecosystem. The few species that remain in any numbers share the plains with the dominant large mammal inhabitants, cattle. Prior to large-scale Anglo introductions, grey wolves, bobcats, prairie dogs, black-footed ferrets, sharp-tailed grouse, coyotes, and even a small population of jaguars co-existed on the plains. The story of the grey wolf (*Canis lupus*), or lobo, as the big plains wolf was called, serves to illustrate the multi-front battle waged against the animals of the plains. Viewed as evil predators, wolves were a source of bounties for cowboys, who often earned more money in the winter months hunting wolves than they would herding cattle for the rest of the year (Doughty and Parmenter, 1989). The federal government also became involved in the extirpation of the grey wolf. The Bureau of Biological Survey (later the US Fish and Wildlife Service) began a predator control programme in 1915 aimed at eradicating wolves (Doughty and Parmenter, 1989). The programme was so successful that grey wolves, counted as "still abundant and very destructive to stock" in 1907 by a government survey, were rated as virtually extinct in Texas by 1921.

Early human history of the Southern High Plains

The Southern High Plains were widely inhabited long before the arrival of European settlers. Following the retreat of the final glaciation, paleolithic hunters inhabited the plains to the eastern slopes of the Rockies. This "thin veneer of lithics" or "paleoindians" (Bamforth, 1988) can be divided into four relatively well-defined divisions within the "Paleo-Indian Period" as recognized on the Southern High Plains: Clovis (11,500–10,000 B.P.), Folsom (sometime between 10,500 and 10,000 B.P.), Plainview (dated to approximately 10,000 B.P. with exact beginning and ending dates unclear), and Firstview (post-Plainview to 8,000 B.P.). These divisions are distinguished by changes in the types of projectile point found in the region and on the plains in general, and in the species of large mammals that these points were apparently used to kill. The oldest certain human occupation known in the study area, as in the rest of North America, dates to the Clovis Period, between 11,500 and 11,000 B.P. These people hunted camel, horse, and several species of elephant (mammoth) found at the Lubbock Lake site in Lubbock County, Texas. Paleo-Indian population densities on the plains probably were much lower than those attained by more recent Native American groups in the region (Bamforth, 1988).

Following the Paleo-Indian or "Big-Game Hunting Period" was the so-called Archaic Period, which Rathjen (1985) dates from about 6,000 to 2,000 B.P. This period refers primarily to a woodland and river-valley culture that subsisted upon small game, fish, and wild plants. A site thought to be representative of this period (age-bracketed between 4,000 B.P. and 1,000 B.P.) in Randall County, Texas, called Little Sunday Canyon (a tributary to Palo Duro Canyon), has yielded stone artifacts, grinding tools, hide scrapers, choppers, and milling stones.

The Woodland tradition emerged around 2,000 B.P. and is signalled by the appearance of pottery derived from the Eastern Woodland. This is the probable time of the introduction of maize agriculture. Around 1,500 B.P., emigrants from the woodlands of the Mississippi Valley and further east brought gardening to the river bottoms of the Missouri and its tributaries. These people and their successors lived in earth lodges or grass-domed houses, growing corn, squash, maize, and tobacco. They travelled on foot and hunted buffalo and antelope, at times driving the animals over bluffs to their deaths. The buffalo was important to their economy, but not critically so (Rathjen, 1985).

Plains Village culture dominated the Southern Plains for 300–500 years. The Plains Village tradition included successful agriculture – specifically the cultivation of maize, beans, and squash. Hunting continued to be an important component. The most important archeological site in the study area is the Antelope Creek Focus, located along the Canadian River and its tributary streams. These ruins are dated from A.D. 1350 to 1450, on the evidence of pottery shards probably derived from trade with the Puebloans to the west (Rathjen, 1985).[2] The ruins once were thought to be evidence of an eastward extension of Puebloan peoples, but recent research contradicts this and has shown the ruins to be more recent. The Antelope Creek people are now thought to have been pottery-making Plains people who first appeared in the central plains shortly after 900 and who gradually migrated into the Canadian River valley about two or three centuries later (Rathjen, 1985: 42). Similarities exist between the Antelope Creek Focus and the Upper Republican culture. Hyde (1959) describes the Upper Republican culture as Caddoan or "primitive Pawnee," from the east. Their pottery was considered similar to that of the "primitive Pawnees" who had come into the High Plains along the Upper Republican River valley in Nebraska at about the same time. These groups of Caddoan origin ("Mound Builders" of the Mississippi) had settled on the headwaters of the Canadian River in the Texas Panhandle after 1200. Archeological dating of pottery shards found in the ruins of adobe and stone slab houses at the Texas Panhandle Pueblo site suggest the ruins are from the period 1300–1500 (Hyde, 1959).

To the south-west, the Puebloans planted corn, tobacco, cotton, squash, and beans. They wove cloth and made pottery, and by 1000 the Puebloan culture was spreading widely throughout the North American South-west. One large ruin attributed to the Puebloan culture was unearthed along the Canadian River in the Texas Panhandle. In later years, the Apaches would drive the Puebloans out of the trans-Pecos area.

Change on the Llano

Some histories of the Southern High Plains suggest that it was only the landscape Anglo-European settlers had to conquer. These accounts imply that no one dwelled on the grasslands; there were only the ground to break, fences to erect, homes to build, and crops to bring in. Clearly, nothing could be further from the truth. The entire area was criss-crossed with people trading, following the buffalo herds, warring with one another, grazing horse herds, and tending fields, people who had been there for more than 300 years, and possibly as much as 1,000, prior to the arrival of Anglo-European settlers. In recent history, the Apaches, Comanches, Kiowas, and Kiowa Apaches all inhabited the Southern High Plains. The earliest European contacts with any tribes on the plains were recorded in 1541 by the Spanish under Francisco Vásquez de Coronado, who travelled from Pecos Pueblo through north-eastern New Mexico, across the northern Texas Panhandle, and into Oklahoma and Kansas. In 1542, according to Hyde's (1959) interpretation of the Coronado documents, the Apaches controlled the southern plains in their entirety from northern Texas up to the Arkansas River. Around this time, Apachean groups apparently occupied the western plains from south-eastern Wyoming down through eastern Colorado and western Kansas into north-western Texas (Hyde, 1959).

Early Spanish descriptions suggest that the first Apache occupants of the region were full-time hunter-gatherers, semi-nomadic groups using dog-pulled travoises to move between seasonal camps. The Spanish adopted the Puebloan Zuñis' identification of these people as *Apaches*, or "enemies." The Apaches called themselves People, *T'Inde*, or by variants of the term *Naizhan*, meaning "Our Kind" (Fehrenbach, 1994: 130–1). Linguistic evidence suggests that the Plains Apaches separated from their northern Athapaskan ancestors (the groups who came over the Siberian land bridge) substantially before white contact, possibly as early as 600. Information gathered by the Spanish from the Puebloans of New Mexico indicates that these Apaches arrived on the Southern High Plains around 1500 (Hyde, 1959). In the sixteenth century, the world of these Native peoples would be irrevocably altered by the arrival of Spanish "explorers" from the south. Alternating between cooperation and trade, and conquest and division, the relationships among the tribes and with the new presence on the plains were complex and fluid. But the Europeans brought horses and guns to the region, and changed the course of history.

By the time of Spanish incursion, Apaches were living as part-time farmers, spending the planting and harvesting seasons in small *rancherías* dispersed throughout their territory and living the rest of the year as nomadic buffalo hunters. They were the first to acquire the horse from the Spanish, but they apparently did not learn horse-breeding as the Comanches did. When sedentary, they lived in adobe houses with flat roofs something akin to those of the Puebloans. Some bands, like the Jicarillas, were advanced in agrarianism; others remained more nomadic.

In 1598, the Spaniard Zaldívar came upon one of the Apache *rancherías*:

> Going on along a large river, evidently the upper Canadian, Zaldívar met more Vaqueros returning to the plains after a trading expedition to Picuris and Taos, where they had exchanged dried meat, tallow, hides, and salt for maize, pottery, cotton cloaks, and small green stones (turquoises). Here Zaldívar saw an Apache *ranchería* or camp of fifty skin tipis colored red and white, "with flaps and openings, all very neat." ... This was in the autumn of 1598. (Hyde, 1959: 9)

By the time the Spaniards were established at Santa Fe (after 1609), the Apaches still dominated the Southern High Plains. The Comanches, who commanded the area when the Euro-American settlers began pouring into the region, did not arrive on the plains until the beginning of the eighteenth century. And although the Comanches and the Utes raided and travelled as far south as the Spanish settlements in New Mexico, they made no attempts to occupy these lands permanently until later. Bamforth (1988) contends that some semblance of tribal locations can be reconstructed *c.* 1650 but more reliable schemas are not possible until after 1700. He cautions that, in considering tribal "territories" on the plains, it is important to remember that the known buffalo-hunting groups in the region were highly mobile. The introduction of the horse obviously increased this mobility, but even the pedestrian groups covered large territories and periodically travelled far outside their usual lands to trade, hunt, and wage war.

Between 1700 and 1725, Shoshonean groups drove the more agrarian Apaches out of the region from the Black Hills south to the Arkansas River. After 1725, these southern Shoshonean groups are referred to as the Comanches, taken from the Ute word *komant*, or "enemy", (written down by the Spanish as *Comantcia*) (Wallace and Hoebel, 1952: 45).[3] By 1775, the Comanches had completely driven the Apaches from the plains, and occupied the Southern High Plains as far south as central Texas and the Rio Grande.

The arrival of Spanish explorers precipitated an era of change on the Llano, both social and environmental. The search for material wealth to send back to their homeland and sponsors spurred the explorers farther and farther into what is now New Mexico and Texas. They brought horses and guns, and eventually cattle, with them. Their very presence in the region disrupted an evolving way of life for the tribespeople living there. The explorers also brought a new dimension to trade on the Llano. And trade broadened networks, widened travel and brought in other ways of life and material culture. It also introduced a military presence in the region which was to lead to sustained conflict between the indigenous peoples and Euro-Americans.

Horses, guns, and trade

The increased migration and interaction with other tribes and Anglo-European settlers and explorers set the stage for the transformation of the physical environment. General tribal migrations continued into the mid-1800s. Modern ethnographers reconstructing tribal distributions place the Comanches in the far south-west, primarily in Texas, with the Kiowas and Kiowa-Apaches just north and east of them in Oklahoma (see figure 2.1). Later changes in territories related to the ability of the various tribes to obtain European horses and guns. A primary cause for uneven access to these resources grew out of the differing policies of the European colonial powers toward trade. Although the Spanish were forbidden to trade guns and ammunition, they did trade other goods, including horses; the French and British (to the east and north) had few horses but were willing to trade guns. According to Bamforth (1988), these trading policies created distinct patterns in the distributions of horses and guns across the plains, with

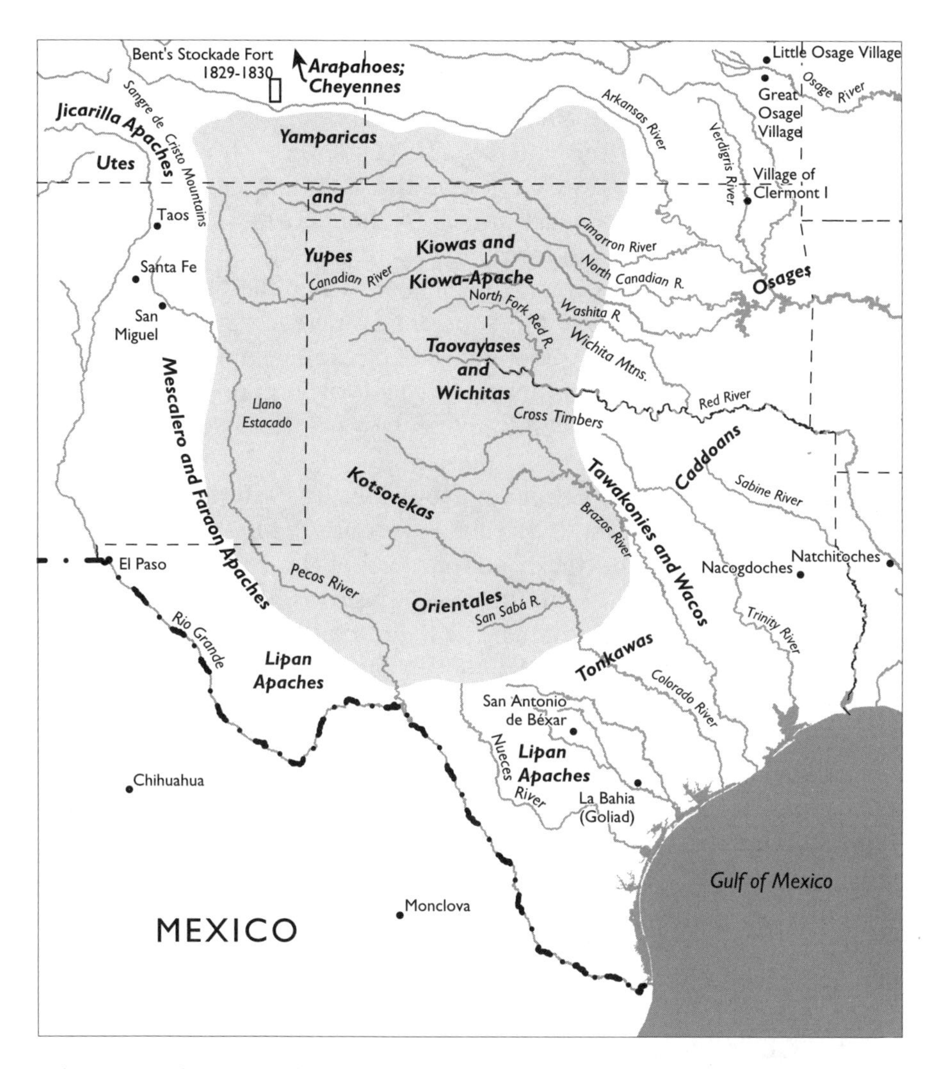

Figure 2.1 **Comanchería and the Southern Plains in the early nineteenth century**
Source: Noyes, 1993: 84.

the numbers of horses increasing toward the south-west and the numbers of guns increasing toward the north-east. The first groups to obtain horses (which were traded earlier than the guns) were those nearest to the Spanish, specifically the tribes in the Chihuahua area of Mexico (Hyde, 1959). They soon learned to ride, and by 1560 these "Mexican Indians" had a good supply of horses (ibid.). By the late 1600s, the Mendoza-López *entrada* into Texas encountered mounted Apaches along the Pecos River.

The Plains Apaches rapidly obtained substantial numbers of horses and developed a style of mounted warfare, giving them a dramatic military advantage over their neighbours. Horses then spread to the Caddoan tribes to the east by 1700 and were obtained in large numbers after 1690 by the Utes and the Comanches. At the same time, these groups distributed horses to their northern relatives; the Shoshones in Montana and Wyoming were raiding the Crow and the Blackfoot on horseback by 1730. The most immediate impact of the adoption of horses was the substantial increase in the area over which these groups could travel, and the advantage gained in hunting the large mammals of the plains, particularly buffalo. With the increase in horse herding, changes in vegetation patterns probably soon followed.

At the same time, the Caddoan groups in the eastern plains began to trade for guns with the French. Around 1740, the Comanches made an alliance with the Caddoan tribes in eastern Nebraska, Oklahoma, and Texas, obtaining for themselves a reliable source of guns. By 1750, every Comanche warrior coming to Taos to trade carried a French musket (Hyde, 1959). The refusal of the Spanish to supply the Apaches with guns even in the face of this situation led to ultimate Comanche dominance.

The advantage created by horses and guns enabled the Comanches to accomplish their conquest of the southern plains in less than fifty years, driving the Plains Apaches south of the Canadian River (in the present-day Texas Panhandle) (Noyes, 1993). The southern plains, which the Spaniards had long considered a part of Apachería, were becoming Comanchería. The Comanches were thus the last "first people" on the Southern High Plains; they were certainly the most renowned, and controlled the Southern High Plains when European scouts, buffalo hidemen, and settlers began trickling into the area.

Comanches and Europeans on the Southern High Plains

The incursion of Europeans onto the Southern High Plains brought about major disruptions in the course of human and environmental history in the region. The Spanish in particular initiated large building projects, erecting missions and governing palaces on a scale not seen before. Large tracts of land were brought under cultivation, with rudimentary irrigation works, and small-scale mining operations were begun in search of elusive lodes of gold. These were unprecedented perturbations on the landscape: until then, agriculture had been limited to small riverine plots, pueblos were low-scaled adobe and stone structures, and pottery had been the primary extractive activity. The Europeans also introduced large-scale herding, breeding horses to support exploration and military expeditions.

Trade networks and militarization of the region

Trade among the native tribes and the settlers to the east and west soon emerged as a major transformational activity that would eventually change the physical and social world of the Llano. In many ways, the trade networks that developed on the Llano and moved across the plains were the start of the transformation of the indigenous ecology. Not only were major trading routes built, with permanent structures, trading posts, and forts to protect them, but foreign materials and animals were also introduced. Guns, in the hands of the tribal peoples and the traders, had a drastic impact on the numbers of wildlife in the area. And both the French and the Spanish were very interested in maintaining or increasing trade with the Native Americans for a number of reasons – to keep peace, garner wealth, and obtain necessities, for example. It has been argued that by the middle of the seventeenth century, the horse trade may have replaced buffalo hunting as the primary mode of production in Comanche life, fully involving Comanches in the developing Euro-American political economy (Foster, 1991). Even food supplies may have been obtained more by trading horses than by hunting or plant-gathering. Although the Spanish attempted to regulate Comanche trade by controlling the frequency and conditions of Pueblo trade fairs, the Comanches remained independent from the Spanish.

The Comanches made their first contacts with Anglo-American traders along the Santa Fe Trail, the 900-mile trail that was built between Independence, Missouri, and the mission of Santa Fe, in the early 1800s. Arguably, the first major environmental disruption in the modern history of the region, the gradual slaughter of the great buffalo herds on the plains, can be traced to the arrival of Anglo traders (figure 2.2). The first traders, the Bent Brothers and Ceran St Vrain, were operating in Comanche country by 1829, and built trading posts for Comanches and Kiowas. In 1833–1834, these traders built a large adobe trading post known as Bent's Fort, on the Arkansas River in what is now south-eastern Colorado. The situation of this post on the Santa Fe Trail made it a rendez-vous for scores of traders, particularly fur traders: the fur trade was at its zenith from 1807 to 1843.[4]

Figure 2.2 **Buffalo herd**
Source: Library of Congress Prints and Photographs Collection.

Comanche trade in horses and mules remained crucial to the New Mexican economy from the mid eighteenth century to the early 1870s and was a significant element of the Texan, Louisianan, and Missourian economies up to the Civil War. Foster (1991) argues that this trade also enabled the Comanches to survive the catastrophic reductions in wildlife and widening land restrictions on the plains. By 1847, the Comanches were experiencing difficulty in finding buffalo and other herd animals. It is estimated that by the 1830s, after the introduction of guns and Anglo traders, the huge southern herd of buffalo was reduced by 20 million to 35 million animals. A Special (Indian) Agent Neighbors reported in 1847 that "buffalo and other game have almost entirely disappeared from our prairies" (cited in Foster, 1991: 47). Some have attributed the disappearance to an underlying ecological cause (e.g., climate change), but most assign the loss to overhunting and land settlement (Richardson, 1933). Findley (1991: 110), writing of the Santa Fe trail, asserts that

> [t]he buffalo hunt as a macho ritual began to draw sporting parties from the East, and the market value of hides and horn brought professional hunters. Indian depredations increased; clashes were numerous, and Congress was pressured to legislate the Indians onto reservations, a job for the Army. By 1867 Kiowa Chief Satanta remarked: "There are no longer any buffaloes around here, nor anything we can kill to live on."

The trade in buffalo hides was highly profitable. In the summer of 1840, fifteen thousand buffalo robes and a large quantity of furs were delivered to St Louis. In the winter of 1841/42, eleven hundred bales of buffalo robes, weighing about 50 tonnes, along with a tonne of beaver pelts, were shipped out of the southern plains. On the trips west, the Bent wagons hauled trading goods and government goods to be distributed to the Native Americans as annuities for ceded land (Gard, 1959).

The sporadic hostilities between Native Americans and trading caravans created a furore, not only about the violence but also because of the threat to the Santa Fe trade. As a result, the governor of Missouri asked for assistance from the military; President Andrew Jackson authorized an infantry escort for the 1829 caravan on the Santa Fe Trail. This introduction of a military presence marked the beginning of the so-called Indian Wars. Skirmishing, mistaken identities, revenge killings, and other violent acts were to plague the efforts of the Native Americans and newcomers to deal with one another for years.

The war with Mexico

Before 1821, Santa Fe was the northernmost provincial capital of New Spain, and Americans who ventured there risked fines, confiscation of goods, and imprisonment. In 1821, Mexico declared its independence and before the year was out, traders from the American frontier were coming in along the Santa Fe Trail. Early in 1823, the Comanche chiefs signed a peace treaty with Mexico, which established rules for trade between the two nations and widened the network of interaction among people in the region.

Comanche trails into Mexico are still visible on the landscape. A few miles north of the Rio Grande, near the present-day city of Del Rio, many trails out of Comanchería converged into a broad roadway, "beaten by the passage of thousands of unshod Comanche ponies" (Fehrenbach, 1994: 251). This trail, called the Comanche Trace, was a distinct landmark running for miles south of the Rio Grande, gradually disappearing as the bands dispersed across northern Mexico. Another trail, farther west, went through the Santa Elena canyon. Other trails ran into the Mexican states of Tamaulipas, Coahuila, Nuevo León, and Chihuahua, over distances of hundreds of miles.

The part of the southern plains now known as Texas had been acquired by the fledgling United States with the Louisiana Purchase in 1803; the territory was subsequently relinquished to Mexico in 1819. The arrival of Anglo settlers in the Texas territory in the 1820s soon led to tensions with the Mexican government. A decade of skirmishing led to the Texas Revolution, launched in October 1835. After gaining independence from Mexico, the Republic of Texas requested annexation by the United States. In 1845, Texas was admitted to the Union, and shortly thereafter a boundary dispute with Mexico resulted in renewed hostilities. The war between Texas (or the United States) and Mexico was fought from 1846 to 1848. The increasing pressure for expansion of US territory in the 1840s, the movement of US citizens into northern Mexico, the political instability of the Mexican government, and financial claims of US citizens against Mexico also accelerated the push to war (Oliva, 1989). The Treaty of Guadalupe Hidalgo, signed in 1848, which detailed the area to be ceded to the United States and fixing the US/Texas boundary with Mexico at the Rio Grande, ended the war. This included the vast region west of Texas to the Pacific Ocean and north to the Oregon territory. The US agreed to pay Mexico US$15 million and to assume

$3.5 million in claims. The war cost the US over 12,000 soldiers (almost 11,000 died of disease) and $100 million. New Mexico became a US territory in 1848.

The Santa Fe trail, the longest of the great trails west, was a major route for military transport of soldiers and supplies during the war with Mexico; the trading route became an important means of fulfilling the expansionist goals of the US government (Oliva, 1989). Native American resistance along the Santa Fe trail increased during this period and the Indian Battalion of Missouri Volunteers under the command of Colonel William Gilpin was stationed along the trail to provide military protection in late 1847 and part of 1848 (Oliva, 1989).

Comanchería in the nineteenth century

The Comanche bands reached their greatest strength during the first half of the nineteenth century as they maintained dominance over the Llano. Fehrenbach (1994) estimates there were around 20,000–30,000 Comanches at this peak. Their territory stretched from the Brazos River to the Arkansas River on the east and north, and to the Sangre de Cristo Mountains and the Pecos River on the west and south. The only real limitation on Comanche activity in the early part of the 1800s was disease. Smallpox, measles, and cholera appeared among the Comanches in 1816 and again in 1839. Whole bands were decimated. Venereal disease was rampant among the mestizo population, and soon spread to the Comanches. Syphilis became epidemic, and by the 1830s and 1840s evidence of European diseases was frequent north of the Red River, all presumably contracted from the Spanish (Fehrenbach, 1994).

Histories of the period suggest that many different Native American groups also were active in this region. The Comanches and the Osages, generally enemies, coexisted peacefully on occasion, for example. There were complicated relationships between particular groups, usually traders, and certain chiefs. Quapaws, Arapahoes, Taovayases, Pawnees, Wichitans, Osages, and others lived around each other or at least crossed each other's pathways. The Comanche–Ute wars had decreased in intensity, in part because the Comanches were raiding in the more profitable areas of Texas and Mexico, rather than westward into New Mexico.

The forced migrations of Native Americans from the eastern United States, which caused such turmoil in the 1830s and led to the death of so many peoples, soon began to have an impact on the activities of the people on the plains. Richardson (1991: 32) writes that "[t]housands of these red men from east of the Mississippi (who were being dispossessed of their lands) were located near the Comanche hunting grounds during the period between 1825 and 1840." The Indian Removal Act, passed in 1830 by the US Congress, compelled thousands of eastern Native Americans to move to what was known as Indian Territory, a region in present-day eastern Oklahoma and Kansas and western Missouri and Arkansas. To protect their territory, the Comanches fought the newcomers at every turn. The US government, fearing that eastern Native Americans would refuse to move into such a hostile region, sent delegations in 1832 and 1833 to establish peace between the various Native American groups, particularly the Comanches and the eastern tribes.

Settlers continued to arrive in Texas during this period, in part because the commercial interests in Texas were advertising in Europe.[5] During the 1840s, thousands of Germans and Irish emigrated to Texas, settling in the north and west, much closer to Comanche territory than previous immigrants (Rollings, 1989). This brought them well within range of Comanche raiding and infringed upon the living space of the Native Americans. Violence spread along the frontier. In the first three years after Texas achieved statehood, 70,000 people arrived, many of them settling in the western part near Comanchería. After Texas gained its independence from Mexico, the Texas government had offered free homesteads of 518 hectares (1,280 acres) to new settlers. Thousands of newcomers, refugees from the general economic depression in the United States at that time, accepted the land offers and began moving west into Comanchería. The Comanches attacked these squatters; the Texans retaliated, forming local military units and groups of "Indian fighters" like the Texas Rangers.

The discovery of gold in California in 1848 brought another flood of Euro-Americans through the region, using the Santa Fe Trail and the Canadian River to make their way west and further challenging the rights of the Comanches to control the plains. These travellers killed Comanche game, used precious wood, water, and grass, and otherwise disrupted Comanche lives (Rollings, 1989). Gold beckoned to a different sort of immigrant: poverty-stricken slum dwellers from the Atlantic coastal cities, as well as fortune-hunters from the farming frontier. They brought disease with them and spread a virulent plague of Asian cholera across the whole expanse of the southern plains. The Comanches, who had kept the Europeans out of their camps, had not understood the importance of typical disease vectors like polluted water, clothing, and other objects any more than had the Europeans. During this period, nearly all of the leaders and many members of the southern Penateka bands were killed by cholera or smallpox (Fehrenbach, 1994).

With statehood for Texas, the US government took on the responsibility of protecting the settlers, the eastern Native Americans who were forced west, and the Mexicans who remained after the 1848 Mexican Cession. The government also inherited the Texans and their running war with the Comanches. In 1849, a line of forts was built across Texas for the purpose of protecting settlers and reducing the conflict. The military presence encouraged European settlers to move further west into Comanche territory. Moreover, for those Native Americans who had never completely reconciled their relationship with the Santa Fe Trail traffic, the newly arriving railroad was terrifying – it now brought hordes of white people through, and to, their lands. In 1830, the entire country had only 117 kilometres (roughly 73 miles) of railroad tracks; ten years later, the track increased to 5,358 kilometres (3,328 miles), then stretched to 14,259 (8,879 miles) in 1850 and 49,324 (30,636 miles) in 1860 – more than in all of Europe at that time (Takaki, 1993: 100). (The railroad ultimately reached Santa Fe in 1880, thus concluding the story of the Santa Fe Trail, the longest-lasting of the great trails west and the site of the most resistance and raiding from Native Americans (Findley, 1991).

By the late 1850s, the Comanches were more or less completely surrounded by soldiers, well-armed settlers, and "Indian fighters," all in the effort to secure ever more land for "settlement." There was little refuge in Comanchería any more; even peaceful Native Americans were indiscriminately attacked by whites who did not know or care about the difference. Because of this, the Bureau of Indian Affairs Agent for Texas, Robert Neighbors, convinced the federal government to provide land for the Native Americans outside of Texas. In 1858, all Comanche bands as well as other Native American groups were to move north of the Red River, which flows east across the Texas Panhandle, becoming the boundary between Texas and Oklahoma. Only 384 Comanches did so (Rollings, 1989). Comanches living outside Texas had to reach an accommodation with the Upper Platte and Upper Arkansas agents to have access to licensed traders, retain some measure of safety for their families, and be relatively free from military interference (Foster, 1991). These agreements were reached between certain band leaders and government officials.

The relationship established between Texans and the Comanches, uneasy at best, was distorted by ignorance and misperception. On the one hand, Texans viewed Native Americans as the opposite of "civilized" Euro-Americans – bent on the "wanton destruction of frontier homesteads." On the other hand, when trade with the Comanches was desirable, those same wanton, barbarous thieves were tolerated and even defended. By the 1840s, however, Comanche *land* was as important to Texans as Comanche trade. The "ruthless Comanche warrior" was a representation useful for justifying both Anglo occupation of Comanche lands and the establishment of an independent Anglo-American government and militia to protect the frontier better. For the Comanches, the loss of wild animal populations meant that raiding and trading were more important than ever, thus ensuring continuing conflict between the Native Americans and the newcomers. As Hagan, in his excellent history of the allotment, or reservation, years, pointed out (1976: 13):

> The Comanches had come to depend upon other items available only from the Comanchero or his eastern counterpart. The hard bread, sugar, and coffee these traders brought to the plains enriched the Comanche diet, and the firearms, kettles, and metal tools simplified the Indian's struggle for existence. The Comanches were nomads, but even the Quahadas had become dependent on trade goods long before they became reservation Indians.

The final act was about to unfold for the Comanches. The pressure to preserve the Llano for settlement dictated the removal of the Comanches. The means were less than creditable, but very effective.

Life on the Llano had changed, slowly at first, but irreversibly. By the middle of the nineteenth century, the Comanches were pushed to the margins of their own lands, of Comanchería. Settlers were moving into the area, and finding vast reaches of what appeared to them to be unused land. The robustness of the ecological foundation was not apparent to these newcomers, much less its complexity. What had taken 100 centuries to evolve was about to be undone in a scant human generation, measurable in decades.

Notes

1. The ranching era complex has been reconstructed as: buffalo grass (first rank), blue grama (second rank), hairy grama (third rank), plains, Reverchon's, and Wright's three-awn and three-awn grama (*Bouteloua* sp.) (fourth rank), and, lastly, 37 other forbs and short grasses (fifth rank) (Tharp, 1952).
2. All dates hereinafter are A.D., unless denoted B.P.

3. The cause of the earliest migrations on the plains, particularly those of the Numic and Athapaskan groups before European contact, are poorly understood. One theory is that the Shoshones, or Comanches, moved south because of pressure from the Black Hills Dakotas. Others have posited that the Comanches were seeking the larger buffalo herds and the Spanish horse herds to the south (Fehrenbach, 1994). Others again counter that, in fact, buffalo herds were smaller in the southern plains (Bamforth, 1988).
4. According to Gard (1959), traders trafficked in alcohol for hides from Native Americans. As a tragic result of this accord, "the tribesmen to satisfy their craving for strong drink, killed very nearly all the buffaloes from certain ranges" (Gard, 1959: 57).
5. This method of advertising, boosterism, is discussed in greater detail in chapter 3 in the context of Anglo settlement patterns in the study area.

3

Early stages of transformation: 1860 to 1900

The decline of Comanchería

"What shall we do with the Indians?" asked a writer for *The Nation* in 1867, as the Irish crews of the Union Pacific and the Chinese crews of the Central Pacific raced to complete the transcontinental railroad. The "highways to the Pacific" must not be obstructed. The Indians must either be "exterminated" or subjected to the "law and habits of industry." Civilizing the Indians, he suggested, would be "the easiest and cheapest as well as the only honorable way of securing peace." This would require the integration of Indians into white society. "We need only treat Indians like men, treat them as we do ourselves, putting on them the same responsibilities, letting them sue and be sued, and taxing them as fast as they settle down and have anything to tax." (Takaki, 1993: 101, quoting Alfred Riggs, "What Shall We Do with the Indians?" *The Nation*, vol. 67 (31 October 1867), p. 356)

The disintegration of the Comanche dominance of the region was matched by the escalation in promotion of the area to new European migrants from the east. The removal of the native tribes was the means of making the Llano available to those who would transform it from a relative wilderness of indigenous plants and animals into the latest extension of the United States. The hegemony of the agrarian ideal was looming large over the course of events on the plains, and the path to environmental transformation seemed inevitable. Of course the two processes – the disintegration of tribal dominance and the escalation of settlement – were driven by the presence of the US military, the hide hunters, and the entrepreneurial and governmental interest in Euro-American settlement of the area. The Civil War created an interruption in the fighting between Native Americans and the newcomers, but by the end of the nineteenth century, the remaining indigenous peoples were on reservations and the railroads were bringing in new settlers by the thousands.

The physical environment suffered from the arrival of European settlers. The wide diversity of wildlife making its home on the Llano was for the most part displaced either by actual extinction or by severe degradation through aggressive predation by the turn of the twentieth century. The region became little more than a vast repository of resources ready for extraction or exploitation: the first commodity that was exploited to the point of extirpation was the buffalo.

By the middle of the nineteenth century, the Comanches were constantly under pressure to give up more of their hunting territories to Anglo settlement (Foster, 1991). Most of the bands based their operations beyond the settled frontier, their territories being roughly as shown in figure 3.1. Within two years of the establishment of the first forts in Texas, in 1849, a new line of forts was built even further west. Pressure on other Native American groups continued to increase as well, with further impacts for the Comanches. In 1855, Cheyenne and Pawnee migrations southward pushed the Comanche bands to below the Upper Arkansas River, where they remained until 1864 (Foster, 1991). The state of Texas, more or less forced by the federal government, created two small reservations for the southern Comanches; some of the Penatekas occupied one for a time, but most refused. Remnants of the Caddoan tribes were put on the other (Noyes, 1993). They were not so involved in the trade for horses and cattle and could not maintain traditional subsistence economies in the face of the major changes in the plains environment, particularly the loss of bison and the loss of rangeland. Also at this time, the Comanches were divided into the upper, or northern, and the lower, or southern, groups for administrative reasons. While the reservations became the base for those so-called lower (or southern) Comanches directly engaged in trade with Anglos, they never attracted more than 450 Comanches at any one time (Foster, 1991).

After the outbreak of the US Civil War in 1861, few troops were available to chase Comanches. According to Fehrenbach, thousands

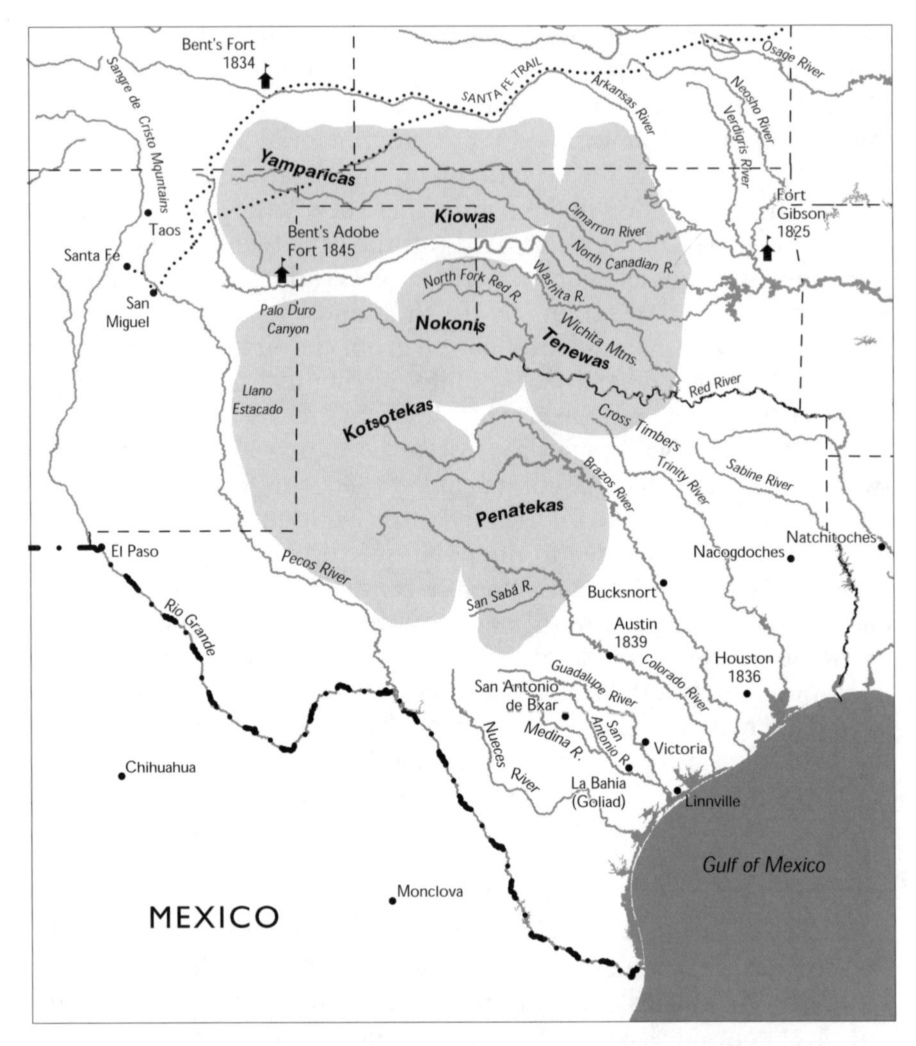

Figure 3.1 **Comanchería and the Southern Plains, *c.* 1845**
Source: Noyes, 1993: 178.

of Euro-American women, children, and old men packed their belongings and abandoned their fields and homes along the frontier. Ten thousand head of cattle were driven off the plains during 1863/64 alone and the frontier in Texas retreated eastward between one and two hundred miles, back to the vicinity of Fredericksburg and Austin (Fehrenbach, 1994). Native Americans, including the Cheyennes, also attacked whites in the Colorado and Kansas settlements. Various territorial and army forces that organized against these more northern peoples pushed many Comanches and Kiowas further south into Texas.

When the Civil War ended, attention again turned to the troubles along the frontier. In 1864, the federal government sent troops into Comanchería – mostly responding to increased Kiowa raiding along the Santa Fe Trail. After the end of the Civil War, the federal government met with numerous representatives of several bands of Comanches, Kiowas, Cheyennes, and Arapahoes on the Little Arkansas River. As in the past, the Native Americans promised to stop raiding or attacking white settlements and the federal government promised trade and protection of Indian Territory (which was growing steadily smaller). For the Comanches, this meant most of the present-day Southern High Plains of Texas and western Oklahoma. In 1867, the federal government sent another peace commission to meet with the same bands. The resultant agreement was the famous Medicine Lodge Treaty, by which the Comanches, except for the Kotsotekas and Quahadas, agreed to renounce their claim to greater Comanchería and to live upon a reservation of some 14,245 square kilometres (about 5,500 square miles) in Indian Territory. Ten Bears, a Yamparika Comanche leader who spoke for the Penateka, Yamparika, and Nokoni divisions, argued eloquently for his people to be left alone (quoted in Fehrenbach, 1994: 474–5):

> My people have never first drawn a bow or fired a gun against the whites. There has been trouble between us, and my young men have danced the war dance. But it was not begun by us. It was you who sent out the first soldier and we who sent out the second ... If the Texans had kept out of our country, there might have been peace. But that which you now say we must live on is too small. The Texans have taken away the places where the grass grew the thickest and the timber was the best. Had we kept that, we might have done the things you ask. But it is too late. The white man has the country which we loved, and we only wish to wander on the prairie until we die.

The reservation, called the Leased District, established between the 98th and 100th meridians, bounded on the north by the Canadian River and on the south by the Red, had been included in the original territory assigned the Choctaws and Chickasaws when they migrated under the removal programme (Hagan, 1976: 16). This small corner of Indian Territory did not include the better hunting lands in Texas, and so most of the Comanches continued to consider all of the land south of the Arkansas River theirs. Only a few bands agreed to move to the new reservation. And as the Comanches of the Llano Estacado (the Quahadas) and the Kotsotekas had no representatives at the commission's meeting, they refused to leave the freedom of the plains for the reservation. (Figure 3.2 depicts a typical Quahada (Comanche) settlement.)

The reservation period

According to Foster (1991), Euro-Americans used reservations as legal vehicles to obtain ownership of large expanses of land in exchange for much smaller areas, as well as a means of restraining the movements and activities of nomadic Native American communities. For Comanches, reservations served a different purpose – at least in the beginning. In the 1850s, the Comanches had used the reservations as protected enclaves to maintain trade relations with Texans and to evade Anglo military pursuit (Foster, 1991). They also used the Indian Territory reserve provided under the 1867 Medicine Lodge Treaty in much the same way, initially. And they received, or at least were promised, semi-annual annuity payments and biweekly rations. But the Office of Indian Affairs, charged with oversight and maintenance of the reservation system, was poorly run and had an abysmal record of providing rations and general sustenance for the people on the reservations. Ultimately, the incompetence of the Office of Indian Affairs led Congress to abolish it and to form a new commission, the Indian Bureau, which had joint jurisdiction with the Interior Department over Native American affairs and appropriations. Congress then offered agency posts to the nominees of various religious denominations. Thus, the Dakotas became the charge of the Episcopalians, and the Comanches were delegated to the Quakers (Fehrenbach, 1994).

An Iowa Quaker, Lawrie Tatum, was appointed in 1869 as the Comanche agent. He believed that compassion and honesty would solve the problems of the Comanches, Kiowas, and Kiowa-Apaches – warrior peoples "habituated to violence" (Hagan, 1976: 59). He was responsible for several thousand Native Americans speaking nine languages and occupying an area the size of Connecticut (Hagan, 1976). Tatum's headquarters were adjacent to Fort Sill (Junction City, Kansas, was the nearest railhead town). He submitted plans for developing the farmland on the reservation and for providing ample food to the "inmates." The bill came to US$200,000 and the Indian Office promptly denied his request, which meant that unless the Comanches could continue to hunt, they would not be able to support

Figure 3.2 **A typical Quahada settlement**
Source: History Study Pictures, Indians of History, Library of Congress Prints and Photographs Collection.

themselves. At this time, only a few Comanches were staying at the agency; life there offered no benefits over the nomadic life on the plains. And it had become obvious to the Comanches that making trouble gained them attention and food: "the tribes that submitted meekly to the whites and never broke treaties were allowed to starve in obscurity, forgotten on their reservations" (Fehrenbach, 1994: 490).

Military intervention and retaliation escalated in the 1870s in response to increased raiding and attacks by Comanches (Hagan, 1976: 71). At this point, some of the remaining Kiowas and Comanches dedicated themselves to fighting to the bitter end. Kicking Bird pronounced that his "heart was a stone; there was no soft spot in it"; Lone Wolf, a fierce Kiowa raider, foretold: "I know that war with Washington means the extinction of my people, but we are driven to it" (Hagan, 1976: 99).

With all of the remaining Comanches on the reservation, this constituted, for the first time in their history, a gathering of the entire Comanche community. The surrender of the Quahadas on 2 June 1875, marked the cessation of Comanche hostilities and the end of their existence as nomadic warriors dominating the southern plains (Hagan, 1976). There were fewer than 1,600 Comanches present (Hagan, 1976: 139).

On the reservation, government agents controlled the supply of nearly all the economic resources Comanches depended on. Rations were distributed at Fort Sill every two weeks or when available until 1879, when the agency relocated to what is today Anadarko, Oklahoma. Winter buffalo hunts provided the Comanches with meat other than beef and brought income through the sale of hides. The last of these hunts occurred in the winter of 1878/79, when the Comanches were unable to track any buffalo.

The buffalo

Buffalo were killed by the million on the Southern High Plains (Gard, 1959). Slaughtered for the hide industry, the massive herds were rapidly depleted, proving that the animals were not after all "innumerable, boundless or infinite" (Worster, 1994: 66). (An approximation of the enormous herds is depicted in figure 2.2, p. 34.) The buffalo of the Texas plains, known as the Southern Herd, numbering in the hundreds of thousands, were effectively exterminated in four years, from 1874 to 1878 (Pool, 1975). Some have suggested that this mindless slaughter was motivated by aims other than the obvious material gain realized from the very profitable hide trade. The buffalo, they argue, were emblematic of the old West, untamed and unusable by Anglo settlers. Buffalo foraging on the range would be replaced by cattle – the new West, as intended by some divine purpose (Doughty and Parmenter, 1989).

The attitude that the Llano was no more than the sum of its environmental resources, which could be put to better purpose by serving economic and agricultural interests than by supporting native species of animals and plants, was complemented by the belief that the vastness of Texas was matched by the inexhaustibility of the animals and birds. The Treaty of Medicine Lodge (1867) did not prohibit whites from hunting on the High Plains, but hunting parties had stayed away until the early 1870s, perhaps daunted by the prospect of encounters with armed and mounted Comanches.[1] With that threat largely eliminated by the US military's intervention, hunting parties, organized around the aim of bagging as many game animals as possible, began to travel regularly through the Texas plains in the late 1800s and early 1900s in search of buffalo, antelope, deer, wolves, even prairie dogs (Doughty and Parmenter, 1989). Birds were also favoured targets, both for the sport and for the feathers and meat.[2] One researcher has concluded that the death toll for animals and birds on the plains alone during the years from the 1870s through the early part of this century numbered around 500 million (B. Lopez, quoted in Worster, 1994).

Worster (1994) contends that killing on such a staggering scale was the inevitable result of the philosophical separation of the animals of the plains from the total ecology of the plains. Animals were not regarded as an integral part of the region, creatures who should be allowed to live on unmolested in their habitats, exploiting and supporting the ecological system surrounding them. Instead, animals were seen as foraging on grasses where cattle should properly be grazing, or attacking valuable range animals, or being less useful than more familiar animals brought with the European settlers from their former homes (Worster, 1994). As a result, animals and birds were pursued with an unprecedented and probably unmatched ferocity, both by individual hunters and bounty-hunters and by those carrying out the mandates of government policies.[3]

When government rations were not adjusted to compensate for the irreplaceable loss of the buffalo, the Comanches tried raising cattle that they obtained from the agents and through their own efforts. Cattle trails between Texas and Kansas went across reservation land,

and Comanches enforced their legal ownership of the reservation by collecting tolls on the herds that crossed their land (Hagan, 1976). Comanches captured strays and as late as the 1890s were known to raid neighbouring Chickasaw herds. They earned "grass money" by leasing their pastures to Texas ranchers beginning in 1885. This arrangement lasted until 1906 and provided each Comanche with a semi-annual payment. The periodic disbursement of grass money attracted a number of licensed Anglo traders to the reservation and also fuelled Comanche gambling activities. They still had not become self-supporting by the end of the reservation period in 1901.

Reservation life successfully broke up the larger residence bands. An even more significant change in Comanche social and economic relations was engineered by the number of white settlers who poured into the area after allotment. In 1878, only 150 to 170 whites resided on the reservation, typically agency personnel and licensed traders and suppliers. By 1890, trespassers were a persistent problem. In the eleven years leading up to allotment (see chapter 4), they became more numerous and more difficult to control, and by the end of the reservation period, Anglos were illegally flooding onto the reservation in anticipation of its opening to homesteading. When the remaining reservation lands were assigned by lottery, the nearly 1,500 Comanches found themselves in the midst of more than 30,000 Anglo homesteaders (Hagan, 1976: 269).

The destruction or marginalization of the Native Americans in the region has been variously interpreted. Notably, recent authors, especially those who have interviewed Comanche and Kiowa descendants, argue that the people saw their loss as understandable within the context of their history. They displaced the Apaches before them, thus they were displaced by the Europeans. Yet there is a certain bitterness about the way, it was done, through treaties and warfare. To Fehrenbach's mind (1994: 371), the treaty process was a sham:

> Few, if any, of the signatories could read the texts of treaties, or even sign them – they gave assent by touching the pen with which the whites signed – and they assented to loose translations. Often, neither party understood the exact territorial definitions or proposed boundaries; American commissioners frequently set boundaries that were never surveyed. Almost all the Plains tribes had the same social and political organization as the Comanches; this meant that every adult warrior must touch the pen before a treaty became binding upon an entire band or tribe. No treaty ever signed with any tribe upon the plains was ever completely valid, from this fact alone – which every American commissioner refused to recognize. This

rendered effective diplomacy with the warrior tribes almost impossible, if not ridiculous. American agents described Cheyenne, Dakota, and Comanche "nations" as determinedly as did the Spanish scribes.

The Comanches did not like to attend councils; tribesmen would speak their minds and the Anglos would proceed as if they had not heard them. They also did not like to attend councils to which their enemies were also invited: "Americans rarely understood why it was unthinkable for Comanches and Apaches to sit together" (Fehrenbach, 1994: 372). Owing in part to these grave indignities, Fehrenbach argues (ibid.), American diplomacy with the Native Americans was more reprehensible than the brutalities inflicted upon them.

Massacre matched massacre; tortures were repaid in displacements and exterminations. The same forces that impelled the modern European nations to world power made the conquest of the American continent inevitable. From the late eighteenth century, most intelligent Europeans understood that the white entry into the Americas spelled at least a cultural death warrant for the natives.... The question facing the United States throughout its advance across the land was never whether the aborigines should be removed as a barrier to its civilization, but only the best means and timetable of the removal. The question was bound up in politics, ethics, morality, and even sentimentality – but it was handled almost throughout with pervasive hypocrisy.

But Foster (1991) argues that there is an important difference in the perspectives of Anglos and Comanches on Comanche history. For example, whites tell the story of the last buffalo hunt with nostalgia and sadness. They saw the native identity as slipping away, almost lost. In contrast, the Comanches tell the story without regret or finality. The story simply tells of an event that happened in the Comanche community at some time in the past. In Anglo representations of Native American peoples,

too often the activity is emphasized over the people who participated in it. Comanches, however, have not forgotten that the maintenance of a traditional community depends on people, not on a fixed set of behaviors, syntactic structures, or food sources. We have largely overlooked the Comanches who surrounded and killed that last, disoriented bison and then went on with their lives. Those lives continued, bison or no, as did the Comanche community they shared. At the time of the last bison hunt (perhaps sometime in the early 1880s) Comanches already were participating in public gatherings of their community different from those of five or ten years before. They were getting on with the business of what they still talk about as "being Comanche." (Foster, 1991: 2–3)

Foster continues (1991: 167), "It has even been asserted that their community ceased to exist once the landscape in which we first encountered them was remade in a European image."

The assertion that the Comanche community ceased to exist once the landscape was changed is quite remarkable within the context of what we mean by critical ecological zones or sustainable development. These were people who completely lost or were forced off their lands and out of their mode of production. Culturally, they have survived, with significant government help. Certainly, the destruction of the buffalo by the white hunters providing hides for the market, as well as by the Native Americans themselves, would have created a critical ecological zone. Perhaps a small change in climate did exacerbate the decline of the great herds; the most powerful source of change, though, was human action undertaken to meet market demands, make money, take land – the same forces that motivate actions by the current populations of the Llano. If there is a lesson to be learned, it is perhaps this: losing control of the land, and one of its primary resources, the buffalo, completely transformed the lives and livelihoods of the Native Americans on the Southern High Plains. Their communities ceased to exist in practical terms once the land was no longer theirs to roam and the buffalo were brought to the edge of extinction. How different will the outcome be for the descendants of the Euro-American settlers even more dependent on what the land can produce and the water can make possible?

Euro-Americans on the Llano

Ranching

With the forcibly decreased presence of the Comanches, the physical environment slowly began to change as a result of Euro-American settlers' hesitant approaches into the region. The dominant mid-nineteenth-century Anglo presence on the plains, though, revolved around ranching. Settlers attempting to move onto the plains often found their claims already leased to the big ranching enterprises: the largest of these was the XIT Ranch at 1,215,000 hectares (three million acres). Ranching grew out of the herding practices of the Comanches and Mexicans after Texan independence. Limited before the Civil War to the area bordered by Mexico, the Gulf of Mexico, the Nueces River valley, and the Edwards Plateau, ranching expanded north to the High Plains after the Civil War ended (Webb,

1931). With the southern states economically devastated, ranchers began to look toward the growing markets for meat and hides in the northern states, and began driving the herds closer to those markets. Within 15 years, and for the decade following that expansion, cattle ranching dominated the plains. The post–Civil War economic boom rapidly integrated the frontier West into the US economy. Responding to the high prices and high demand for beef from Midwestern and Eastern markets, Texas cattle numbers nearly doubled in 16 years, from fewer than 5 million cattle in 1875 to more than 9.8 million in 1891, an inventory peak that would not be exceeded until 1945 (Texas Department of Agriculture, 1990: 5).

The boom

Colonel Charles Goodnight, thought to be the first rancher on the Southern High Plains, brought cattle from Colorado in 1877 and established the JA Ranch in what is now Deaf Smith county. By 1883, between 25 and 30 ranches were operating in the northern part of the Llano (Sheffy, 1963: 13), including enormous ranches owned by Midwestern and British interests. The Capitol Freehold Ranch, or Capitol Syndicate Ranch (XIT), was formed in 1881 when the Texas State legislature created a 1,215,00-hectare (three-million-acre) reservation in nine counties in Panhandle/Plains Texas (Dallam, Hartley, Oldham, Deaf Smith, Parmer, Castro, Bailey, Lamb, and Hockley). This land grant was used as payment for the construction of a new capitol building in Austin. A group of Chicago businessmen won the bid, built the capitol, and stocked the ranch with over 160,000 head of south Texas cattle. The Matador Land and Cattle Company was probably the second largest ranching operation on the High Plains. The company, financed by Scottish investors, purchased nearly 220,000 hectares (540,000 acres) on the Southern High Plains and leased another 81,000 hectares (about 200,000 acres) during the 1880s.

Although these companies purchased (or traded for) their land tracts, until the early 1880s, most ranchers simply used the free range to stock cattle. Texas, as part of its agreement to join the United States in 1845, maintained public lands, and permitted ranchers to use public domain lands as open range for cattle ranching. Known as range rights, this gave ranchers authority to use the surrounding range and its water without cost (Webb, 1931). Within only a few years, however, greater numbers of homesteaders and cattle on the range meant that competition for land increased, and establishing tenure became necessary. During the early 1880s, ranchers began fencing in their land (and public land they did not own) with barbed wire, sometimes encircling homesteaders (Dale, 1930). By 1883, competition for land between ranchers and homesteaders was so fierce that the Texas legislature had to step in to settle land disputes. Free-range ranching quickly came to an end when most public and school lands were sold and private land was fenced. Fencing was used not only to establish rights to land, but is credited with improving ranching by lowering operating costs, improving the quality of cattle, particularly through controlled breeding, and with settling unwatered portions of the Southern High Plains (Dale, 1930). The availability of cheap, practical, barbed-wire fencing "revolutionized ranching [but] it did not destroy it and would not have threatened it seriously had it not been for its [beneficial] effects on the farmer's frontier" (Webb, 1931: 316).

The number of Texas cattle continued to increase through the 1870s and 1880s and exceeded 9 million head by 1888. Until 1890, most cattle were sent to market through trailing. A series of cattle trails crossed the Great Plains leading to "cowtowns," particularly in Kansas (such as Abilene and Dodge City), where cattle were sold and shipped to Kansas City, St Louis, and Chicago for slaughter or finishing. Most Texas ranchers hired drovers, or trailing contractors, to drive their cattle northward to the Kansas cowtowns (Skaggs, 1973). Through the 1870s and 1880s, cattle trailing to Kansas was less expensive than paying rail freight charges (Skaggs, 1973). About a million head per year were trailed from Texas to northern markets during the 1870s and 1880s (Dale, 1930). Most Texas ranchers drove their cattle either north-west across the Pecos Trail (the Goodnight–Loving Trail), which led from San Angelo to Santa Fe, or north across the Chisholm, Matamoros, and Western trails. North-west Texas and Southern High Plains ranchers drove their cattle along the Potter–Bacon Trail, which crossed the High Plains and joined the Santa Fe railroad in south-east Colorado.

The bust

Trailing was short-lived: the extension of railroads, fencing that broke up the open range, and quarantines to contain Texas fever in the 1880s led to its demise. Texas fever is a cattle disease caused by a microscopic parasite that uses ticks as its vector. Texas Longhorns in south Texas developed immunity to the disease, but thousands of cattle of other breeds died wherever Longhorns were trailed. As early as 1882, High Plains ranchers, led by Charles Goodnight, posted armed men on the Potter–Bacon trail to stop the trailing of south Texas Longhorns across the High Plains (Skaggs, 1973). Kansas reacted even more strongly in 1884 by forbidding the entrance of south Texas cattle into the state except during the winter months, when cattle were not killed by the disease. Quarantines moved Texans to call for a national trail to be built and maintained by the federal government, but few other constituencies supported the effort. The end of trailing finally came in 1889, when the Secretary of Agriculture, J.M. Rusk, quarantined all or parts of 15 southern states. Texans were forced to patronize railroads. Ironically, it was soon after the 1880s quarantines that government researchers at the US Department of Agriculture discovered the cause of Texas fever and showed that it could be eradicated by dipping cattle in a cleansing solution that killed the disease-bearing ticks (Skaggs, 1973). By this time, however, the trails were closing, and few lasted past 1890. Railroads had extended even to the High Plains; a line connecting Fort Worth and Denver was operational by 1890 (Dale, 1973).

While the quarantines on Texas cattle dampened the cattle industry boom somewhat for ranchers on the Southern High Plains, it was a disastrous coincidence of environmental and economic conditions that led to the complete collapse of the cattle trade in the late 1880s. Ominously foreshadowing the coming Dust Bowl calamity by roughly 50 years, the booming cattle ranching on the Plains soon began to have serious impacts on the rangelands. Overstocking the range was leading to heavily overgrazed land. The range was viewed as an endless supply of free grass, and more and more cattle were brought onto it; there were 7.5 million head on the Great Plains by the mid-1880s (White, 1991: 223). With unusually mild winters, these growing numbers of cattle were surviving, and increasing grazing pressure on the range each subsequent season. The most heavily grazed grasses soon declined and hardier, woodier plants gained dominance. But these plants were much less edible, and with the return of very harsh weather in the winter of 1885 the poorly fed cattle could not survive. On the Southern High Plains, before the worst overgrazing was occurring, one steer could thrive on five acres (2 hectares) of rangeland; by 1880, it took 50 acres (20 hectares) (White, 1991). The onset of freezing temperatures and high winds in 1885, combined with the loss of range due to overgrazing, brought heavy losses. The herds on the Southern High Plains were decimated.

The Northern High Plains underwent the same disastrous conditions the following year, and the combined losses contributed to the worsening economic scene. A national depression was deepening, and nervous creditors, made more concerned by the cattle losses, soon began calling in their loans. To satisfy debts, ranchers began selling off their herds, driving prices down even further. Ranchers who could moved further west to the deceptively sparse grasslands of the Great Basin and the intermontane west. The cattle ranching that remained on the plains was just a remnant of the flourishing cattle trade of the late 1870s and early 1880s, never to recapture the teeming numbers of that era.

Settlement

At the same time as the ranching expansion on the plains, several attempts were being made by emigrants to the plains to establish farms and communities. These early attempts, in the 1870s, were founded along the breaks, off the eastern edge of the Caprock, below the plains in the small canyons and headlands of the few small streams. Access to water was the most important factor in locating a farm, and surface water was very scarce up on the Llano. The first farming settlement established on the Llano, the town of Estacado, was started in 1878 by a community of Quakers from Indiana. Of the four families who moved to the new town of Estacado that year, only one remained after the first winter. Paris Cox and his family planted a variety of crops the following spring and saw good harvests. The news of good weather and yields enticed other members of the Indiana group out to the plains and within ten years there were 23 farms around Estacado (Green, 1973).

The arrival of the railroads and the coincidence of cyclically driven favourable weather conditions increased the movement to the plains in the last two decades of the nineteenth century. When the harsh climate and drought forced people off the plains, railroads and news of good rainfall and harvests brought them back in droves. Clearly, people wanted to believe that the plains could be made habitable and even prosperous. Vast amounts of land were available and the hope that these were good arable lands was being made reality by the will that they would be in the near future.

Settlers poured onto the plains in large numbers in the early 1880s.

Reconstructed climatic data show that the weather was highly variable throughout that decade, with more rainfall than was typical for the region for several years in a row (Lawson and Stockton, 1981). But the latter half of the 1880s brought a turn toward much drier than normal weather, and farmers began pulling up stakes and abandoning homesteads as crops failed again and again. Dust storms blew up in the spring of 1889 and raged for days. By 1891, the plains were rapidly depopulating. In Floyd County, for example, there were 176 farms at the beginning of 1891. By the end of 1892, fifty-five farms remained. Farmers also began sinking into debt as they sought mortgages to make up for the loss of crop income (Green, 1973).

Legislation, railroads, and boosterism

The further transformation of the Llano was under way. The extreme hardships of homesteading notwithstanding, people still arrived on the plains to stake claims. Several interests were at work promoting the transformation of the plains into farming communities. The most prominent of these were the federal government, acting through federal land laws (which in turn influenced the state land laws), the railroad companies, and the individual boosters and promoters of the nascent towns of the plains.

Legislation

Land laws were an early impediment to successful farming on the plains, although the intent of the laws was always to promote settlement. In distributing the public domain, the federal government devised land laws directly aimed at promoting a certain type of settlement and development. The original land distribution scheme, the Land Ordinance of 1785, provided for large tracts of land 9,654 metres by 9,654 metres (six miles by six miles). These tracts were then subdivided into sections measuring 1,609 metres (one mile) to a side, which were in turn again divided into four quarter-sections and then sold. Small landholders would best represent the sort of nation-building envisioned by the founding fathers, it was believed. Numerous small tracts of land farmed by individual families promoted the ideals of independence and self-sufficiency so embedded in the notion of Americanism.

The reality of land distribution fell far short of these lofty goals. To fund public works, the cash-poor federal government began issuing land scrip that could be bought and sold. This led to land speculation and the amassing of large holdings by purchasers with no intent to settle on those lands (White, 1991). What White refers to as extra-legal means to acquire land were in fact adjustments to the federal land distribution system necessitated by Major Powell's "fact of aridity." It soon became obvious that roughly 65 hectares (160 acres), a quarter section, of semi-arid plains could not support a farmer, let alone a farmer and a family: "without irrigation ... [the] quarter-section farm ... was not a ticket to independence but to starvation" (White, 1991: 142–3). The land laws that settled the eastern woodlands and the prairies would not lead to permanent settlement of the semi-arid, and often arid, plains. In response the federal government, unwilling to change the land distribution system, moved to change the climate.

Congress legislated the Timber Culture Act (1873), which was intended to foster the growth of trees on the plains, grounded in the popular theory that rain would increase in areas where the trees were planted and the sod broken up and ploughed (White, 1991). When the climate did not respond, Congress then passed the Desert Land Act (1876) allowing settlers to acquire a full section (256 hectares or 640 acres), with the provision that the land be brought under irrigation within three years of filing a claim to the land. However, in regions of negligible surface water, three years was scarcely a reasonable deadline to "prove up" a claim.

In the midst of these developments in the schemes to settle the plains, John Wesley Powell issued his *Report on the lands of the arid regions of the United States* (Powell, 1879). Recognizing the necessity of access to water for survival on the semi-arid plains – 508 millimetres (20 inches) of rainfall annually was seen as the minimum required for self-sustaining agriculture – Powell created a plan for allocating the lands beyond the 100th meridian, the geographic demarcation of sustainable agriculture (Powell, 1879). The basic framework of the plan entailed creating "hydrographic basins", units of land centred around their relation to available water (Powell, 1890). While Powell clearly did not comprehend the availability of the extensive stores of water underground, or foresee how access to that water could make much of the arid lands irrigable, he clearly did understand that, given semi-arid to arid conditions, "values inhere in the water, not in land; the land without the water is without value"

(Powell, 1890). Therefore, Powell's scheme called for the inclusion of access to water in all land grants, attaching water rights to the land, creating irrigation and pasturage districts, with cooperatives instituted to manage irrigation systems and rangelands (White, 1991).

Powell's plan was rejected by the US Congress, which chose to continue pursuing individual land grants. Attempts to modify land distribution were a de facto acknowledgment that land laws needed to be different in the semi-arid and arid lands, but actual restructuring of land laws to reflect the climatic reality of the West seemed unlikely. Law must follow custom, Powell advised, where existing law does not address the physical reality (Powell, 1890). In Texas, where land distribution schemes had their beginnings in the early days of Mexican control, landholdings were structured to accommodate the arid and semi-arid climate. The large holdings amassed by ranchers reflected this Mexican influence, overriding the federal tendency to limit individuals' ability to own too much land. Texas did not institute any form of homestead act; only a relatively small percentage of land was ever given away, and that chiefly for the purposes of building schools, with some land going to veterans, immigrants, and the railroads. A total of about 21 million hectares (52 million acres) was provided for schools, nearly a third of the land area of the state (Webb, 1931). Most of the agricultural land in the state had been distributed by 1880 (climate notwithstanding); what remained was largely without access to surface water. A scheme was devised and legislated providing that the land be classified as agricultural, grazing, or woodland. Grazing land was further divided into watered or dry land. Homesteaders could then apply for a single section of agricultural land, two sections of grazing land, or a single section of each.

Although this system acknowledged the difficulty of maintaining a living in the semi-arid and arid parts of the state, like the Llano, the law was still amended almost annually to allow settlers to obtain ever larger tracts of land, particularly unwatered land. By the early years of the twentieth century, settlers in the Llano were eligible for holdings of up to 2,048 hectares (5,120 acres), eight sections of land (Webb, 1931). Thus the Texas laws, despite starting from a position that recognized the near-impossibility of surviving on the standard quarter-section of the humid prairies and eastern lands, also evolved to accommodate the unforgiving climate of the Llano and other parts of the state.

Railroads

A second major force behind the settlement of the plains was the railroads. If it had been at all possible for the railroads to have induced more rain to fall on the plains, the railroad companies would have ensured it. As it was, railroad companies were avid promoters of settlement on the plains. They planned to build lines in advance of settlement, to reap the profits of flourishing markets, should they arise. Therefore, the railroad builders were determined to do everything possible to ensure that settlement would occur. Towns along the lines were necessary to supply the engines with fuel and water, and to fill the freight cars with goods, farm products, and cattle.

Railroad companies began promotional campaigns to lure settlers to the plains: they published brochures, magazines and pamphlets extolling the virtues of the region, hired agricultural and migration specialists, held fairs, and ran demonstration and model farms (Blodgett, 1988). Specific regions or types of agriculture were featured in brochures such as the Fort Worth and Denver City Railroad Company's booklet on Texas "with Special Information Concerning the Panhandle," "The Panhandle and South Plains of Texas," published by the Atchison, Topeka and Santa Fe Railroad Company, "Farmers Make Good in the Panhandle and South Plains," from the Santa Fe Company, and the Rock Island Line's brochure entitled "Panhandle Country for Beef Cattle and Dairy Farming" (Blodgett, 1988).

The railroad companies published testimonials as to the great variety of agriculture that could be practised on the plains as well as various attestations to the success of farmers already in the region. For credibility, however, the railroads also recognized the importance of addressing the issue of rainfall directly. Perhaps in the interests of honesty, much, though by no means all, of the promotional literature turned away from the standard propaganda on how the rails or steel ploughs increased rainfall, or how increasing population increased rainfall; they instead focused attention on the great potential in dry-land farming and irrigation (Blodgett, 1988).

Additionally, railroad companies' brochures, in which actual US Weather Bureau precipitation tables were reprinted, always included statements from experts confirming that the timing of rainfall was actually more important than the amount. Therefore, although farmers from more humid regions might look askance at rainfall averages of 355–500 millimetres (14 to less than 20 inches) annually, they were to be reassured that the rains fell reliably when needed – in the winter and early spring for good planting conditions and then again during the growing season, to promote high yields and abundant harvests (Blodgett, 1988). The railroads' record in expanding settlement on the plains was at best mixed. Maps of planned railroad lines from the 1870s and 1880s were far more optimistic than what was ultimately built. Today, the Llano is served chiefly by freight lines with very limited passenger rail service.

Boosterism

Boosterism, the third major force for settlement on the Llano, is promotion of a place through paid advertising in the form of booklets, pamphlets, postcards, and in at least one notable instance, a motion picture. Two basic approaches were used: first, that the climate was actually very favourable to agriculture, with the correct adaptations and technological advances; second, that the plains were the future, a healthful, industrious place to relocate, raise families, and found prosperous communities.

The first approach, misrepresenting the unpredictable climate and the cyclical drought, had been a mainstay of the railroad companies' advertising. But the boosters took it far beyond that. An entire sub-industry emerged in the late 1800s to contradict eyewitnesses' accounts of their failed experiences on the plains. Whole publications appeared, dedicated to the premise that irrigation would soon turn semi-arid lands into bountiful farms. *Irrigation Age*, started by the editor of the *Omaha Bee*, William E. Smythe, was foremost among them. "The Greatest Irrigated Farm in the World" (illustrated), "Remarkable Growths Under Irrigation," and "Prosperity in Irrigation" are the titles of a few representative articles published in *the Irrigation Age* in the early 1890s (*Irrigation Age*, 1891–1893). A tireless promoter, Smythe crusaded throughout the western states, lecturing and hosting congresses on the possibilities of irrigated agriculture. Smythe would later, in 1911, write a book extolling the changes brought about by irrigation, optimistically titled *The conquest of arid America*.

The difficulty in gaining access to water to irrigate with was not addressed in the promotional literature of the late 1800s. Shallow groundwater could be reached by windmill-driven pumps but could be quickly tapped out; what was necessary for agriculture was simply more rainfall. Numerous books were written on the scientific advances that proved how amenable the region really was to climatic change – how the incomparably rich soils of the plains needed only to be broken up to begin soaking up available moisture like a sponge. "Rainfall follows the plough" was a popular motto, shorthand for a scientific theory of the power of steel ploughs to attract rainfall (Smythe, 1911). Even the advance of the railroads was purported to increase rainfall (American News Co., 1874).

With the realization that drought was as likely as normal rainfall (and frequently more likely), new interest was generated in so-called dryland farming. The promise of dryland farming was touted as the new way to institute agriculture permanently on the plains. J.A. Widtsoe described (and promoted) techniques recommended by the self-proclaimed evangelist of dryland farming, H.W. Campbell, in his 1911 treatise on the subject (Widtsoe, 1911). These techniques recognized the importance of retaining as much soil moisture as possible. But Campbell also preached deep-ploughing of fields, compacting subsoils and loosening topsoils through the windy summer and winter months. The adoption of this method inevitably led to severe wind erosion in dry years (Kraenzel, 1955). But in wetter years, as the latter half of the 1890s were, the increased rainfall and the dry-farming methods led to an era of great prosperity on the plains. This was clearly the most effective boosterism of all.

A secondary theme of the promotional campaign to populate the plains with farmers and town builders centred on the healthy environment and the limitless potential for wealth and success. These messages had strong allure for Europeans, easterners trapped in cramped cities, and for farmers feeling crowded on the prairies. Blodgett (1988: 86) describes the phenomenon:

> The work of the railways and promoters was significantly enhanced by the continuing demand of restless America for new lands, especially agricultural lands.... [T]hese lands held special attractions for those in search of new and cheaper farms, for those in need of healthier climates, for ethnic and religious groups seeking new communities, and for individuals desiring new homes and opportunities.

With increasing settlement and coincident favourable weather, land prices began to rise from US$10 to $30 per 0.4 hectare (about one acre) in the 1880s to $80 to $125 per 0.4 hectare in the 1890s (Blodgett, 1988). Even this was useful to the boosters – here was evidence of how valuable the land actually was, and how it gained in value with settlement. And, the message became, one had better act fast, before the land on the Llano was as expensive as the prairie lands. Land agents, having learned the lessons of the drought-driven exodus, also were seeking prosperous buyers who would not move away, leaving mortgages and debt behind them. Thus, land agents began promoting the potential to improve upon one's situation, to start again in a place where a family could have an influence on the kind of community it would live amongst. Numbers of churches, schools, and the lack of saloons and depravities were underscored. Agents sent photographs back east of "healthy children, tree-and-flower encircled homes, and if possible, prosperous-looking businesses and church buildings" (Blodgett, 1988: 89). The Industrial Survey of the city and county of Lubbock, compiled by the Lubbock Chamber of Commerce, included this promise in its introduction:

> Any prospective settler coming to Lubbock and upon arrival here finding the average conditions different from those shown herein will be refunded total expenses to and from Lubbock. This guarantee is given by reason of our desire to present only the true circumstances surrounding our city and county (Lubbock Chamber of Commerce, 1915: 1).

Those circumstances included ample provision "with the greatest of all necessities – water," with approximately 70 per cent of the purported 24 inches [600 mm] of annual rainfall coming during the growing season" (Lubbock Chamber of Commerce, 1915). Again, the timing of rainfall was emphasized to allay the concerns of eastern farmers: "This amount of rainfall is ... equal to 35 to 40 inches [890–1,000 mm] of rainfall annually in a rolling tight land country" (Lubbock Chamber of Commerce, 1915).

Not only was there reported to be an inexhaustible shallow water supply of "soft and clear water" and "rich chocolate loam" waiting "in perfect readiness the coming of the man with the plow and the hoe," but this was a "country of life giving sunshine and pure air and water, where natural conditions tend to favor and prolong life" (Dillard-Powell Land Co., 1908/1909). East Texans, the targeted audience for this particular booklet, would experience a new ease of living in an extremely healthy climate with pleasant and mild conditions (Dillard-Powell Land Co., 1908/1909). Another booklet, distributed by the nascent Lubbock County Chamber of Commerce, focused on the purity and healthfulness of the air, water, climate, and region in general (Lubbock Commercial Club, 1909). Also noteworthy was the ease of making a profit with agriculture and ranching. A popular feature of most booster literature were the testimonials of people who arrived on the plains poor and soon became wealthy beyond their wildest dreams.

Conclusion

A picture emerges of the people who migrated to the Southern High Plains. There was the population targeted by the more refined efforts of boosters, for one example: these were the conservative families, already successful farmers, who knew what they wanted in their communities, the kind of people they wanted as neighbours. "Come and get a home among people who are liberal hospitable charitable law-abiding, and where peace and harmony abound ..." (Dillard-Powell Land Co., 1908/1909). The authors frequently added, "and white." Race was not subtly addressed: several tracts referred to the absence of African-Americans on the south plains and the "benefits" to be derived from this (Dillard-Powell Land Co., 1908/1909)). Studies of the racial makeup of the western states found similar "enclavement": African-Americans, Chinese, Hispanics all were attracted to specific areas, separated from the larger Anglo settlements (Deutsch, 1992).

With the Native Americans contained and off the plains, settlers attempted to farm the Llano. The vagaries of the climate were soon revealed, as drought and severe winters forced people off the Llano. Boosters promoting the region spoke to certain traits in settlers: individualism, self-reliance, stubbornness, and optimism. While it probably did not require extraordinary vision to see wheat where grass was growing, it did require some fortitude to keep that vision through years of drought, wind, broiling summers, and harsh winters. Perhaps because there was so little "new" land left to claim in the West, people became determined to build homes here on the Llano. There really was no place left to go to begin again. And, for that first generation of pioneering farmers who stayed on the plains, the rewards were just a few short years away.

Notes

1. Rathjen (1973, 155) notes that there was much confusion over the Medicine Lodge Treaty. He doubts that Native Americans were aware that the treaty did not explicitly prohibit white hunting and did not obligate the US army to prevent white intrusion. Since Comanche leaders argued publicly that the High Plains should not be hunted by whites, it is doubtful that leaders willingly and knowingly signed over rights to the land. The leader of the first large hunting party, J. Wright

Mooar, was also confused over the status of the High Plains, and went, in 1873, to the commanding officer at Fort Dodge to seek permission to hunt on the High Plains. The commander could not prevent the hunting party from entering the High Plains.

2. These outings were often sponsored by the wildlife agents in the region, reflecting the vision, most government officials held regarding their charges. The earliest state office for conservation in Texas, that of Fish Commissioner, held its mission to be the introduction of carp into Texas rivers and lakes (Doughty and Parmenter, 1989). The National Audubon Society pushed the state to legislate the first protective measures for birds in 1903, and with growing awareness of the ongoing extirpation of several species, a game agency was added to the Fish and Oyster Commission, so renamed after its reorganization in 1895 (Doughty and Parmenter, 1989).

   The aim of this agency, however, was the preservation of game for hunting. Not until the 1920s did the agency move to regulate hunting by requiring licenses and bag limits for hunters in Texas. The state even began to move toward restocking species (deer, antelope, and turkeys, at first) in regions where they had been extirpated (Doughty and Parmenter, 1989). These efforts, albeit aimed at preserving hunting in Texas, at least signalled an awareness that conditions were irrevocably changing, and that the resources of the Llano, and the rest of the West, were indeed exhaustible.

3. Other species considerably less dangerous to stock were also targeted, including coyotes and prairie dogs (*Cynomys ludovicionus*) – the latter because of the belief that their burrows could be responsible for broken fetlocks and falls for horses and range animals. Brought to the brink of extinction on the plains by extensive poisoning programmes, prairie dog populations have rebounded slightly, and along with other species that have been reintroduced to the plains, are no longer the target of extermination programmes. This was due in part to a growing awareness in the 1960s of the reckless approach to wildlife in the United States. In 1966, the Endangered Species Act was passed, and then updated and strengthened in 1973 and again in more recent years. It is clear, however, that some animals are truly extinct and others so attenuated that they may also disappear; the disappearance of any semblance of the original, seemingly inexhaustible communities of animals and birds attests to the sweeping influence of Anglo settlement on the Llano.

4

The new century and the new ecology of the Llano: 1900 to 1945

Replacement

The beginning of the twentieth century was the start of a new era on the Llano. A fortuitous turn of wetter than normal weather, a growing trend toward mechanization, and motorized farm equipment were inducing settlers to farm larger and larger tracts of land. And with the Comanches all but removed from their lands, particularly after the allotment or privatization of the reservations, the Anglo-European settlers had only to prove up their claims to realize boosters' promises of profitable and prosperous farms. All the success of the first few decades of the twentieth century obscured the subtle signs of the disaster to come, however, as fragile lands, uncertain markets, and unpredictable weather combined to undo the new ecology of the Llano. The ecological transformation of the Llano, begun in the few preceding decades, was now accelerating.

The Llano Estacado was one of the last regions of the conterminous United States to be populated by Anglo settlers. Emigrants from the east and the more settled Midwest passed through the semi-arid and subhumid shortgrass prairies on their way west; there was little in the forbidding and sere landscape to capture the imagination. The flat featureless Llano contrasted greatly with the rolling plains of

Iowa and Missouri and eastern Nebraska and Kansas, cut through with rivers and streams. The lack of timber to clear was an asset to some, but the fearsome reputation of the Comanches and the absence of surface water were clearly reasons to keep going.

Surely one major reason why settlers ignored the potential of the high plains grasslands was the general belief, promoted through school books and popular gazetteers, in the existence of a Great American Desert in the interior of the United States. This had been first reported by Major Stephen H. Long as he explored the West for the United States government (1819–1821), and maps were soon drawn featuring a great expanse between the prairies and the Rocky Mountains labelled the Great American Desert. Dendrochronitic climate reconstructions have shown that in the long-term cycles of drought and non-drought, the era of a great expansion of westward settlement, the 1820s through the 1840s, were indeed a time of even lower than normal rainfall (Lawson and Stockton, 1981). The typically brown and yellow grasses of the southern plains were most likely even browner and sparser than in years of normal rainfall.

The myth of the desert, never very widely nor deeply held, was deliberately replaced in the "romantic" histories of the plains by promoters who, one imagines, were battling to counter actual settlers' reports of failure, drought, and starvation. Bowden (1975) points out that, in effect, the myth of the great American desert was a necessary condition for the further settling of the plains. By nurturing a belief in the great difficulty of surviving in the region, those who were surviving there became plainsmen and women writ large. There also was then an opportunity to show how the ingenuity of the American farmers was serving to create a garden in the desert, with the resulting effect of luring more and more would-be farmers into the region.

The post-allotment Comanche community, 1901–1941

The mere presence of the Comanches, albeit in attenuated numbers, was still considered an impediment to settlement. At the beginning of the reservation period, the land occupied by Comanches was not considered desirable farmland because of its poor-quality soil. By the early 1880s, though, the Anglo-American desire for more farmland brought pressure on the federal government to open up large tracts of Native American land to homesteaders. Under the Allotment Act of 1887, communal lands of the tribes were divided and distributed as private property; the Comanches and Kiowas lost their common reservation. Under the Dawes Act, passed by Congress in 1887, the federal government was empowered to purchase reservation lands from the various Native American tribes, leaving each tribal member with a 65-hectare (160-acre) allotment and opening the millions of hectares that remained to Anglo homesteaders. In January 1890, Commissioner of Indian Affairs T.J. Morgan made the following announcement (cited in Hagan, 1976: 201):

> The 8th of February, the day upon which the "Dawes Bill" was signed by the President and became a law, is worthy of being observed in all Indian schools as the possible turning point in Indian history, the point at which the Indians may strike out from tribal and reservation life and enter American citizenship and nationality.

"Franchise Day" was to be observed by songs, recitations, and so forth, to convey the meaning of the allotment law clearly. It was not popular among the Comanches and Kiowas: Chief Lone Wolf voiced his opposition immediately upon hearing of the Act. But apparently only one organization working for Native Americans' welfare, the National Indian Defense Association, opposed allotment (Hagan, 1976). Ultimately, the Comanches conceded but Quanah Parker (the renowned Quahada chief), Tabananaka, Eschiti, Cheevers, and Howeah, all eminent tribal leaders, made it clear they wished to proceed slowly. The Cherokee Commission was set up in the 1890s to negotiate the value of the reservation lands and the implementation of allotment with the tribal peoples. When the Commission visited Fort Sill and Anadarko they told the Native Americans, "[n]ow you have an opportunity to sell to the Great Father all the land you cannot use for homes for his white children" (Hagan, 1976: 205).

Congress did not authorize allotment to begin among the Kiowas, Comanches, and Apaches until 1900. The 65-hectare allotment was inadequate for self-sufficiency in a region with less than 500 millimetres (20 inches) of rainfall per year. Because at least six hectares (15 acres) were necessary to support one cow, it would be impossible for the people to become self-supporting even with cattle raising on their individual allotments of 65 hectares (Hagan, 1976: 207).

The Treaty of Medicine Lodge (1867), which created the original Comanche reservation, required that any sale of land by the three tribes would have to be approved by three-fourths of the adult males. Therefore, the Cherokee Commission would need at least 422 signatures, given that there were some 550 adult males on the reservation at that time, and very few men had indicated a willingness to sell their common land (Hagan, 1976: 210). A number of influential Native Americans and whites worked behind the scenes, however, and the signatures started flowing in. The Kiowas were more reluctant. Nonetheless, the commissioners obtained the requisite signatures despite rumours of people wishing to retract their decisions to sign. Ultimately, the Commission threatened the Native Americans with the possibility of receiving even less acreage. According to Hagan, accession to the treaty doomed the Comanches and Kiowas to the poverty that some of the Kiowa headmen had foreseen. Allotment not only provided more land for white settlers, it also served to fragment and disperse Native American communities. The belief was that if the community's cultural practices, such as common property ownership and group-based religion, could be destroyed and the Native Americans made to live as whites, their assimilation would be complete.

In choosing their 65-hectare sections in 1900 and 1901, the Comanches had little idea of what their future lives would be like. They chose locations along stream beds for access to water, wood, and shelter from the open plains; they tended to camp or choose allotments near other kin or family, or friends. These choices were guided by traditional notions of the physical landscape and of private relationships.

After allotment, Comanches were subject to the same agricultural uncertainties as the Anglo farmers, but they were not able to compete as effectively owing to a lack of capital. Only 50 Kiowas, Comanches, and Kiowa-Apaches were farmers in 1925, and 559 in 1931 (out of total populations of 5,023 in 1925 and 5,500 in 1931) (Foster, 1991: 102). Their cattle herds had also declined. By 1910 only 2 per cent of the members of the three tribes were engaged in stock-raising: "Anglo thieves, Anglo laws that held Comanches responsible for the damage their cattle did to crops, and the end of federal rations in 1901 all contributed to the exhaustion of Comanche base herds" (ibid.).

The boom years on the Llano

The turn of the century was the start of an era of great prosperity on the Llano for the Euro-American settlers. Settlement boomed. Eight hundred and three farms were counted in the 1900 agriculture census.[1] By 1910, there were 2,001 (there was a total of 417,770 farms in Texas that year); in ten years the number more than doubled to 4,089 (compared to 436,033 farms state-wide in 1920). And by 1925, there was a nearly eightfold increase in the number of farms in 1900 to 6,161 farms established on the Llano. Most of the farms by 1925 were large – between 40 and 200 hectares (100–500 acres). Of 2,445,554 hectares (6,042,880 acres) total area of the region, 1,188,669 hectares (2,937,161 acres) were farmland.[2] This represented anywhere from about 20 per cent of a given county's total area to nearly 80 per cent. Cropland actually harvested (in the previous year) was somewhat less: 462,447 hectares (1,142,692 acres) (US Department of Commerce, 1925–1935).

The dramatic increase in numbers of farms did not go unnoted. A 1909 report of the director of the Texas Agricultural Extension recommended that of four proposed new Agricultural Experiment Stations, two be located west of the 98th meridian. Two of the five proposed sites for the new stations were on the Llano. Ultimately, three new stations were opened west of the 98th meridian with one on the Llano, in Lubbock (Texas Department of Agriculture, 1909).[3] The process of transforming the Plains into valuable productive farmland, the veritable Eden promised by 1880s boosters, was under way.

Several factors contributed to this rapid growth. Weather favourable to agriculture was the most important. Records show unprecedented rainfall amounts nearly every year between 1900 and 1929 (Lawson and Stockton, 1981). Cities were growing and so was the demand for agricultural commodities. But the confluence of two major events changed the trajectory of the Plains and set the region on an immutable course, ultimately leading to the second complete transformation of the Plains in the American mind. The first of these events was the invention and mass production of mechanized farm equipment; the second was World War I (1914–1918).

Motorized tractors, first appearing in the 1910s, were ideally suited to farming on the plains. Level land, stretching as far as one could see in all directions, was the perfect setting for mechanical tractors. Larger and larger tracts of land could be brought into production and harvested with fewer and fewer people – exactly the land and labour conditions of the southern plains. In fact, the Llano was one region of the High Plains that did not follow the pattern of decreasing numbers of larger and larger farms establishing itself throughout the prairie plains and the Midwest. There were still few enough settlers on the vast Llano so that farms could be very large without consolidating existing farms (White, 1991). Farmers invested heavily in the new equipment, leading to increasingly heavy debt burdens, which necessitated putting more acreage under cultivation.[4]

This marked the development of heavily capitalized, technology-dependent agriculture on the plains. Farms became "production sites" in some instances, with the owners living elsewhere, usually in the cities, and coming onto the land for a few days to plant crops and then again to harvest them (White, 1991). Such "suitcase" farming meant a fundamental change in the relationship between the farmer and the land. The intimate knowledge of the land and the weather so identified with pioneering farmers was replaced with an agricultural version of the assembly line, with even farmers working at a remove from their means of production and more intent on the added-value end products. Even for the majority of farmers who did maintain farm homesteads, the scale of operation had changed, and the aim was to maximize profits in order to perpetuate the acreage the equipment made possible, and to keep up the payments on the equipment essential to keeping the acreages in production.

World War I ratcheted up the production levels in the grain belts. When European markets lost access to Russian wheat owing to the Turkish blockade, the Europeans turned for the first time to the United States to import grain (Worster, 1979). The demand more than doubled the going price for wheat, the principal crop on the Llano. As the war intensified, European demand increased and it became a farmer's patriotic duty to plough under more grassland to plant in wheat. By 1919, land under cultivation on the High Plains had increased to 5,463,450 hectares (13.5 million acres); 4,451,700 hectares (11 million acres) of that had been in native grasses (Worster, 1979). The federal government had passed wartime food production measures, including the Food Control Act of 1917, guaranteeing a set price (at least US$2.00) per bushel of wheat, further integrating farmers into an international production system, guaranteeing a price, and cementing an incentive to keep production levels high and increasing, if at all possible.

A thriving cotton culture also grew up within the region during the early part of the century. The first successful cotton crop was produced on the Llano in 1903, with the first gin for processing the lint opening in 1905 (Belt, 1983). By the end of the 1920s, farmers in a number of counties on the Llano Estacado devoted more than 50 per cent of the total cultivated land to cotton (Gibson, 1932). Farms were much larger in size than those of the older Cotton Belt (in eastern

Texas, Louisiana, Mississippi, and other areas of the South), and the large, level fields required much less labour for production. A US Department of Agriculture study conducted from 1924 to 1927 found that it took about 96 hours of labour to produce a hectare of cotton in the Lubbock area compared to 380 hours in North Carolina (Browne, 1937). As a result, the net production costs were comparably lower as well. A Lubbock Chamber of Commerce survey of the area's cotton industry in 1924 found that 60 per cent of the improved land was in cotton and a total of 121 gins were in operation. Land was still available at that time and the belief was widespread that a farmer could pay for his farm with one cotton crop (*Lubbock Morning Avalanche*, 21 September 1924, p. 1, cited in Gordon, 1961). In 1930, the region became one of the top cotton-producing areas in the nation, with a peak crop of one million bales (Belt, 1983). It would be over a decade before that yield would be approached again.

Economics drove the transition from ranching to cotton that occurred during the early decades of the twentieth century: a bale of cotton produced on half a hectare could bring in US$50, the price of four or five yearlings, which might require 20 times the area for grazing (Gordon, 1961). Cotton also brought cash into the largely barter economy of the time. By 1958, approximately 17 per cent of the cotton grown in the United States was produced on the Llano Estacado (ibid.). This percentage increased to 30 by 1980 (Bednarz and Ethridge, 1990).

The "cotton culture" on the Llano was built from the tenacity and creativity of the farmers, and the educational and economic institutions that supported (and continue to support) agriculture within the region. It was unclear to many growers in the early days of farming on the Llano Estacado whether cotton would be as drought-resistant as the grain sorghum that was grown for feed. According to Gordon (1961), the Texas Agricultural Experiment Station at Lubbock was largely responsible for doing studies that proved the comparable status of cotton vis-à-vis the time-tested sorghums. Don L. Jones, noted agronomist and superintendent of the Experiment Station at Lubbock from 1917 to 1957, outlined five factors explaining the success of cotton in the study area. These were: good soil; introduction of the row-crop tractor; development of a cotton plant that could do well with a relatively short growing season, scant rainfall, and high winds; the presence of groundwater for irrigation; and the self-reliance and adaptability of the people who settled the area (Jones, 1959). Other factors would have to include the availability of credit to finance irrigation, land and machinery, and yearly inputs; and the government's special commodity farm programmes. Without the last two in particular, it would have been impossible to maintain cotton production in the Llano Estacado.

Complementing the trend toward greater acreages under cultivation were the ongoing rapid innovations in farm equipment. Some 200 different companies, including Ford Motor Co., McCormick-Dressing and John Deere, were manufacturing tractors by 1917 and creating new equipment, like the one-way disc plough, which soon became essential to modern farming (Worster, 1979). It is important to keep in mind that these were good-weather years (a tendency toward drought in the mid 1910s notwithstanding) and H.W. Campbell's dryland farming methods were producing large yields and good harvests. The one-way disc plough was designed to break up the surface soil layer into a fine "crumb," as Campbell advised, in order to absorb any available rainfall more completely. Farmers following the recommended practices would plough up the surface of their fields after every rain, supposedly to trap the moisture in the soil (Worster, 1979; Widtsoe, 1911).

The stage was being set through cultivation practices, market demands, capitalization demands, and technological innovation for either ever-increasing massively scaled agriculture, or an environmental and economic disaster on a massive scale. Writing at the very end of the boom era, one researcher remarked that "[t]he large scale system of agriculture now in vogue on the High Plains is producing and maintaining a thriving civilization with a newness and color all its own" (Gibson, 1932: 11). By the late 1920s, the grains–cotton vegetative complex was a successful and permanent replacement ecology for the shortgrass and scrub complex. Dryland farming methods, in conjunction with higher than normal rainfall amounts, were producing high yields (Browne, 1937). As Worster (1979) describes it, nothing changed during the boom years; created conditions merely accelerated trends already established. The myth of the garden had successfully supplanted the myth of the desert at last through the perseverance and work of everyday men and women.

The bust

Of course, neither myth reflected the truth, although the spectre of the Great American Desert was about to loom large over the entire southern plains region once again. With the start of the new decade came the end of the good weather. The southern plains were stunned by a sudden onset of drought in the early 1930s. Droughts had come and gone before on the plains – small-scale farmers lost everything, dusted out, migrated east or further west, but on a small scale. With large farms, heavy debt burdens, overextended, poorly managed lands, and no rain, farmers seemed destined to lose on a big scale. Evidence shows that the drought of the 1930s was not the severest on record.[5] In fact, this drought is not even ranked among the five worst droughts since the early eighteenth century, as detected through tree-ring data (Worster, 1994: 101). However, the consequences of the drought and the concurrent international depression were the economic and ecological devastation on the plains.

What made the Dust Bowl era one of the most disastrous episodes in the history of the United States? If it was not the result of the worst drought on record, why were the results of the drought the worst in living or recorded memory? The answer lies in that confluence of the several major developments in agriculture and the markets for agricultural products in the 1910s and 1920s, and in the widely held denial of the climatic reality of the region. In some ways it can be argued that expectations, both presumed and acted on, caused the Dust Bowl.

When the US stock market crashed on 24 October 1929, western farmers did not immediately sense that there would be any impacts on their livelihoods or the economic health of their communities. There was a belief that because the West was built on actual resources, grain and fibre in particular, speculation and investment practices back east were irrelevant. In reality, the West would suffer more and longer, for the markets for the commodities the western half of the United States produced had collapsed.

On the southern plains and the Llano specifically, the immediate response to the commodities market collapse was to plant more wheat, creating a glut and driving prices down even further. The crisis deepened over the next few years, with increasing numbers of farmers sinking into debt as prices for farm products kept falling. Foreclosure rates soared as farmers were unable to keep the notes current on their equipment and land. Farm values dropped anywhere from 25 per cent to 45 per cent between 1930 and 1935 (US Department of Commerce, 1925–1935).[6]

Things were bad for cotton producers as well. High production had suppressed prices at a time of declining demand, following World War I. In 1926, the price of cotton dropped dramatically. A poor

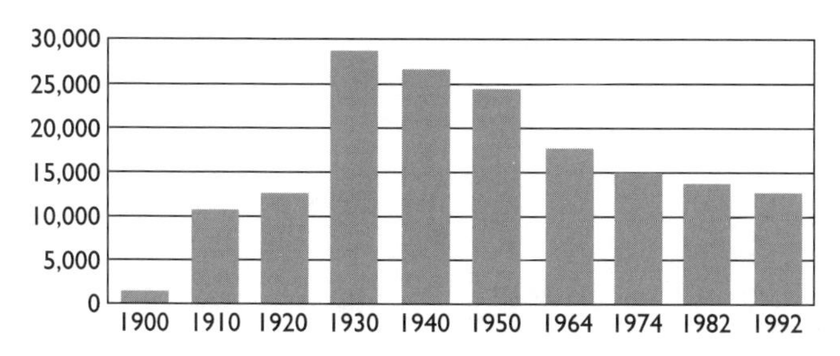

Figure 4.1 **Number of farms (1900–1992)**

season resulted in crop losses of over 60 per cent in some communities (Gordon, 1961). The cotton farmers joined the appeal for relief assistance from the federal government. After a banner year in 1930, a large number of growers voluntarily curtailed their planting; the deepening drought of 1934 reduced output even further. When the drought continued into the early part of 1935, it was clear that it was not just another "dry spell." These bleak years brought to an end the initial phase of cotton culture development on the Llano Estacado – a phase in which the individual was pitted against the uncertainties of climate and competitive market (Gordon, 1961). With the exception of the late 1960s, the number of farms in the area has been decreasing ever since this period (figure 4.1). And, although population increased between 1930 and 1940 throughout the region as a whole, nearly half of the counties lost population during the 1930s (US Department of Commerce, 1900–1940).

Relief programmes aimed at the farmers were sparse and ineffectual prior to Franklin D. Roosevelt's election to the presidency in 1932. Beginning in 1933, sweeping relief programmes were instituted with the goal of getting cash and food into the hands of desperate farmers. But for the farmers of the plains the worst was still to come. The spring of 1935 brought no rains, and in April the dust storms began. The legendary storm of 14 April 1935 (Black Sunday) ushered in an era of unimagined ecological devastation for the southern plains. Enormous clouds of black dust from the plains blew across the continent, drifting on the streets and buildings of Washington DC and New York City; even ships at sea found the fine black dust settling on the decks (figure 4.2). Drought had recurred throughout the known history of the Llano; this time, though, when the drought returned, millions of hectares had been stripped of the indigenous grass cover, and were now exposed to the stiff winds that blew every spring. With

Figure 4.2 **Dust storms over Texas Panhandle, March 1936**
Source: Library of Congress Prints and Photographs Collection. Photograph by Arthur Rothstein, 1936.

the added impacts of the drought, the result was staggering. Some have estimated that 13,350,000 hectares (33,375,000 acres) were stripped of their grass cover and left vulnerable to severe erosion in that spring of 1935 (Worster, 1979). A contemporary account (Great Plains Drought Area Committee, 1936: 8) reported:

> The dust storms of 1934 and 1935 have been visible evidence to nearly every American living east of the Rockies that something is seriously wrong. The extent of erosion on the Great Plains has not yet been accurately measured. It is safe to say that 80 percent is now in some stage of erosion.

Furthermore, the blowing dust and erosion was "a situation that will not by any possibility cure itself. A series of wet years might postpone the destructive process, yet in the end, by raising false hopes and by encouraging renewal of mistaken agricultural practices, might accelerate it" (ibid.).

Survival

Intervention on many levels was necessary to stop the erosion, to provide food and shelter to the thousands of dusted-out residents, and to preserve agriculture in the region. Federal programmes and intervention on a grand scale seemed essential. By merging recovery efforts, several conservation programmes were begun that provided jobs and sought to mitigate the environmental destruction brought about by overproduction and poor farming practices. The Shelterbelt Program was one of these. Its aim was to plant a shelterbelt of trees from the Dakotas to Texas to help ease wind erosion on the plains while employing thousands of displaced farm workers and labourers. Although the programme was somewhat successful in parts of the Midwest, there was simply not enough rainfall to sustain tree growth in the Dust Bowl counties (Worster, 1979).

Another early programme, the Drought Relief Service (1934), also focused on repairing the environmental damage resulting from overgrazing. Millions of cattle were bought by the federal government and moved off the range or killed – 25 per cent of the cattle bought were on the Texas plains (White, 1991). A payment system was also devised to reimburse ranchers for reduced grazing and better water management. This programme was not without strings; the Taylor Grazing Act was also passed that same year (1934). This Act required that the grazing lands in the public domain be under federal control, administered by local boards, under the assumption that if overgrazing could be controlled, more grass could be grown and the erosion problems would also naturally be controlled. How much of the plains should be returned to grassland for grazing purposes was subject to debate. Memories of the 1880s cattle market crash and the increasing glut of cattle raised questions about the wisdom of creating more rangeland.

Recovery

Permanent retirement of farmland and its reversion to grassland (an early precursor to current calls for a buffalo commons on the plains) was a major aspect of the Great Plains Drought Area Committee's plan for recovery of economic and environmental health on the Plains. This committee, appointed by President Franklin D. Roosevelt, comprised the heads of various resource and agricultural agencies, among them Hugh H. Bennet, the chief of the Soil Conservation Service, Harry L. Hopkins, Administrator of the Works Progress Administration, Henry A. Wallace, Secretary of Agriculture, and the chairman of the Rural Electrification Administration, Morris L. Cook. Reporting directly to President Roosevelt, the Committee called for voluntary retirement of "submarginal" farmlands and replanting them with grass or trees. The most severely damaged lands (estimated at about 15 per cent of the total area) would be brought under government control. And the government would retain the authority to restrict or prohibit certain practices by farmers on other lands if those practices were considered to be potentially damaging (Great Plains Drought Area Committee, 1936). Most importantly, though, the Committee stated (ibid.):

> The fundamental purpose of any worthwhile program must be to not depopulate the region but to make it permanently habitable. Any other outcome would be a national failure which would have its effects, tangible and intangible, far beyond the affected area.

The actual recovery programmes fell far short of the goal of permanent reversion of farmland. The notion of such clear government intrusion into everyday farming practices was not welcomed, despite the helping hand the government was also extending. The emphasis shifted to temporary retirement. Under the Agricultural Adjustment Act of 1933, the Secretary of Agriculture was given wide authority to raise agricultural prices in a variety of ways, the chief means being payments to farmers who reduced their farm production, and secondarily, to create marketing agreements for producers with direct oversight by the US Department of Agriculture (Hansen, 1991). Paying farmers not to grow crops for which there was no market was an effort to reduce surpluses and bring land out of production. For farmers, this was a radically different approach to farming, and it made many of them uneasy; it soon became obvious, however, that it was saving many farms from economic devastation. The Agricultural Adjustment Act, characterized as a programme of planned scarcity, was amended several times throughout the decade, and stands as the foundation of farm programmes even to the present day (White, 1991).

Short-term retirement of farmlands was not solving the wide-scale environmental problems, either. Recognizing that, the USDA moved ahead with plans for land stabilization and created the Soil Conservation Service with the Soil Conservation and Domestic Allotment Act of 1936. Moving onto the windblasted wheatfields of the plains, federal experts introduced contour ploughing, terracing, and other soil-conserving measures, meeting with some success in stabilizing the dunefields so prevalent by the mid to late 1930s (figure 4.3). Another successful offshoot of the Soil Conservation Service was the establishment of soil conservation districts. In the recognition that farmers would be more likely to follow locally organized programmes, soil conservation districts were organized at the grassroots level. Using labour from the Civilian Conservation Corps and from local farmers employed through the Works Progress Administration, soil management techniques were demonstrated across the plains. Although this represented a breakthrough in the political management of the environment of the plains, the districts, as they were slowly formed, "were a limited and illusory evidence of progress toward Great Plains reform – not true land-use planning groups, but only mechanisms to promote the agronomists' tools, methods, and narrow view of the land" (Worster, 1979: 220).

Irrigation was believed to be the solution to recurrent droughts, but capital for investing in irrigation works was exceedingly scarce for farmers who were barely surviving the depression and the weather. Irrigation was not new to the Llano Estacado; the cattle ranchers had used wells coupled with windmills since the 1880s. By 1900, the existence of subsurface water in the area was widely known and was being mapped by the US Geological Survey. As Green (1973) said, the "search for water" was largely a search for the economically feasible technology to bring the water to the surface in quantities suffi-

Figure 4.3 **Abandoned farm in the vicinity of Dalhart, Texas, June 1938**
Source: Library of Congress Prints and Photographs Collection. Photograph by Dorothea Lange, 1938.

cient for irrigation purposes. Not surprisingly, irrigation gained in popularity in the 1930s and 1940s. The 1935 Census of Agriculture counted 2,916 irrigated farms in the Llano Estacado. These farms were largely within the areas of shallow water. Truck farmers in New Mexico were irrigating and, in Texas, cotton was the most important irrigated crop because good profits could be realized despite the low prices.

It has been estimated that half a billion dollars in relief funds were sent to the Dust Bowl region during the 1930s. Some portion of that was in direct cash relief; the remainder was used to create programmes and agencies to support farmers over the long term. A major initiative coming out of the 1930s depression and farm crisis was the Farm Security Administration.

Begun in 1937 in direct response to the uprooting of tenant farmers and the devastation of small farmers, the Farm Security Administration sought to stabilize the small farmer and keep the family intact on the farm. One goal was to cut surpluses by reducing cash cropping and turning acreages back to subsistence farming. An education effort was organized to teach farm families how to grow the food they needed and rely less on the cash economy (USDA, 1941). The major market for American farm products was Europe, and European farmers were rebuilding their own farm economies, resulting in decreasing demand for those products. The Farm Security Administration created a series of complementary programmes that sent Administration workers into the farmers' homes to outline strategies for becoming solvent and self-sufficient. They met with mixed success. The aims of the Farm Security Administration programme embodied the ideal of the yeoman farmer held dear from the earliest days of westward expansion. A patchwork quilt of small farms spread across the plains, all independent and engaged in self-sufficient agriculture, was seen as the optimal path to national as well as personal security, a guarantee against the despair and crushing poverty all too evident in the late 1930s in the agricultural sector of the United States.

As the plains slowly emerged from underneath the metres-deep drifts of sand and dust, with the return of the rains in the early 1940s, several factors leading to the environmental disaster became more obvious. Not only were the farming practices disastrous, overriding economic and global political conditions also reinforced the trend to careless and unconsidered farming. Competitive farming led to the consolidation of farmlands, resulting both in huge unprotected tracts of ploughed land, but also in increasing tenancy. Many people, if not most people, fleeing the Dust Bowl were "tractored out," not necessarily "dusted out." As farmlands were bought up to maintain profit levels against falling grain prices (the answer was always to put more land under cultivation), those farmers who could not compete lost their lands to suitcase farmers. Tenants were also the first to go when troubled economic times arose. There was more money to be made if the farmer could tend all the land himself without renters. One man with a tractor could cover more land in a day than any number of tenants; labour costs fell and maintenance costs for equipment were always lower than the costs of maintaining farm labourers. Any of these trends without the potential for serious drought in any given year probably could have been successfully modified and continued. But the fact of water scarcity, in light of the highly variable climate, was undeniable.

In the end, what lifted the High Plains out of the Dust Bowl and the Depression were institutional arrangements that made temporary relief programmes into permanent farm support programmes, and, most importantly, the discovery of an apparently inexhaustible water supply to replace the farmer's eternal dependence on rain. The ultimate toll of the Dust Bowl years is incalculable: 800 million metric tonnes (850 million tons) of earth blown away each year; 900 metric tonnes per cultivated hectare (408 tons per acre); 12 centimetres of topsoil blown off 4 million hectares, and at least 6 centimetres lost off another 5.5 million hectares, all by 1938 (Worster, 1979: 29). At any given time during the period, one-fourth to one-half of all farmers in the region were on relief, employed through the Works Progress Administration or getting cash payments or low-interest loans. A total of 3.5 million people had left the Dust Bowl region by the end of the 1930s, 2.5 million of them farmers (West, 1990).

The data speak volumes, but the photographs of Dorothea Lange, the songs of Woody Guthrie, and the books by John Steinbeck and Paul Taylor revealed the human cost that cannot be captured by numbers. Hopes and dreams were destroyed along with the topsoil. Health was ruined along with cropland. Dignity and independence disappeared with the emergence of the overpowering need for assistance and charity. A way of life was very nearly lost; for those forced to migrate, that way of life was lost forever.

Native Americans, displaced from their lands, were also subject to the economic and environmental disasters of the 1930s. The Comanches earned some income from leasing allotments to Anglo farmers, but with downturns in agricultural prices those leases became worthless. In addition to lease money and per capita payments from the sales of the allotment lands (194,260 hectares, or 480,000 acres, along the Red River left over from joint ownership), some capital was obtained through employment in the Anglo economy. But out of 4,683 Kiowas, Comanches, and Apaches in 1934, only 15 had full-time jobs (Foster, 1991: 105). Many, perhaps one-fourth, received government relief. Outright sales of allotments (perhaps the intent of the Allotment Act all along) then became a means of survival for many families. Some claim that sales of allotments were actively encouraged by the local Indian agency, and by 1940, some 65,000 hectares (160,000 acres) out of a total of slightly more than 218,500 hectares (540,000 acres) had been sold (Foster, 1991).

Moving ahead

The history of the Llano has been marked by boom years and busted lives. As the "Last West," the region seemed destined to be the repository for all the remnants of frontier dreams. Sturdy pioneers struggling to make new lives against occasionally overwhelming odds represented the idealization of the American experience. Even in the face of the complete destruction brought by the raging dust storms, that ethos seemed indomitable. Without irony, the Lubbock Chamber of Commerce applied in 1934 to the Texas Rural Homes Foundation to create a farming community in the vicinity of Lubbock. Its main purpose? To offer to as many as 100 of the 1,000 families on relief in the city of Lubbock the opportunity "to return to rural life and start farming thus developing income, reducing relief rolls, and producing food and fiber," and becoming "self-supporting" (Lubbock Chamber of Commerce, 1934). The reality of the environmental conditions was, nevertheless, undeniable. As the region slid into drought conditions with an unpredictable but ominous regularity, it was perceived as a challenge to the character and strength of will of the settlers. Those who managed to stay on did well in good years, and therefore were in a better position to take advantage of government programmes and to persevere through the bad times.

But the survival of a few provides the material of lore to pull others to the region. And, gradually, an identity is created, that of a people who defied the odds and prevailed. Thus, in remaking the Llano, the settlers remade themselves. With the descent into the Dust Bowl, they were in danger of losing that sense of identity. The federal-level response recognized this (Great Plains Drought Area Committee, 1936):

> The nation has profited by the courage and endurance of the people of the Plains. We have all had large responsibility for the direction of the settlement and for the development of agricultural conditions in the area. We cannot discharge ourselves of the obligation thus incurred until we have helped them create, within the natural and climatic conditions which can be prepared against but cannot be controlled, a secure and prosperous agriculture.

But, in the end, the Llano remained a region in search of a foundation to support the economy built there. Without the water more humid regions take for granted, agriculture and the way of life dependent on it are uncertain. The troubled decade of the 1930s, and the hard times brought about by overproduction and misguided dreams, left their mark. Physical scars on the land served to remind people of how treacherous the climate could be. Episodic drought came to be an expected companion to life on the Llano. Single bad years such as 1917, a year of very low rainfall, came and went. But the prosperity brought about by robust markets and pressing demands for what the plains can produce under the best of circumstances guided beliefs about the ever-possible potential for the plains, as always, in "next year country."

Notes

1. Data are for Bailey, Cochran, Crosby, Floyd, Hale, Hockley, Lamb, Lubbock, Parmer, and Potter counties.
2. Again, these data are from the ten counties listed in note 1 above.
3. The Station was opened shortly after the report was issued. A short-lived drought that year was so severe that all of the demonstration crops failed.
4. White (1991) reports that the new farm machinery displaced so many horses that the French government sent cavalry officers to the western United States to purchase surplus horses for military use during World War I.
5. Recent (1996–1997) drought conditions on the southern plains are considered far more extreme than conditions during the 1930s, for example. By mid-May, 1996, crops on 4,451,700 hectares (11 million acres) had been lost (*Wall Street Journal*, 22 May 1996).
6. These figures are given for the average percentage decrease in total farm value for the ten representative Llano counties listed above.

5

Expansion and exploitation: 1945 to 1980

Recovery and maintenance

The post–World War II era was a time of tremendous growth in agricultural production, standard of living, and population on the Llano. The industrial boom, spurred by wartime production levels and fuelled by a vigorous domestic economy, also supported widespread investment in technological innovation, and its subsequent implementation in the farmers' fields. Farmers on the Llano benefited from all of these factors, resulting in an unprecedented era of growth and prosperity for the region. Early on, in the early 1950s, it did appear that a permanent, inexhaustible supply of water was beneath everyone's feet just waiting to be used. The first great wave of irrigation brought incredible success. In true homesteading fashion, everyone rushed to stake their claim to the immense supply. Perhaps in an effort to "prove up" that claim, water was pumped at a furious rate for the first twenty or so postwar years. As wells began to dry up and drillers went deeper and deeper, some began to acknowledge that there might in fact be a limit to the water, economically or physically.

The chief responses to the dust storms, drought, and misery of the

1930s were institutionalized conservation practices, farm subsidy and support programmes, and the search for a reliable source of water. The last of these – having adequate water – was believed to be the underlying solution to all the problems of settling and maintaining agriculture on the Llano. As Webb (1931) tells the story, "[t]his would be a fine country if we just had water," the new settler is reported to have said to a dusted-out farmer. "Yes," comes the reply, "so would hell." In "next year" country, an adequate rainfall in the spring after decent snowfall in the winter is always the solution to the current year's drought (Worster, 1979). The irrefutable truth that every dry spell ended with a rain drove the farmer to hold out if at all possible for one more year. Finding a means of providing a steady and reliable water source would surely resolve once and for all the question of the suitability of the plains for livelihoods and communities built on agriculture. And it was yet to be proven that cotton culture and ranching were feasible long-term options for the Llano.

Water

That water existed under the sod at some distance was widely known. Windmills dotted the plains, drawing water for watering tanks on the range and supplying the farmhouses' needs. In 1898/99, Willard D. Johnson, a geologist with the United States Geological Survey, compiled a detailed analysis of the climate, vegetation, soils, topography, and water resources of the High Plains. His report, *The High Plains and their utilization*, described with surprising accuracy the extent of groundwater reserves on the plains. Johnson (1900–1901, 646) also responded to the Irrigation Age enthusiasts:

> There is a prevalent belief that means will be devised by which, acre for acre, this great accumulation of water within the ground may be put to irrigation use on the surface. At the same time, a notion, finding its expression in the generally accepted term "underflow" and involving serious error so far as contemplated results are concerned, prevails, also to the effect that groundwater has lateral motion fairly comparable to that of surface water – i.e. that it is continually renewed at such rate that it could be inexhaustible as a supply for irrigation.

However, Johnson explained (1900–1901: 653), the potential usefulness of the groundwater on the plains was confounded by the nature of the resource:

> The volume of loss which the body of the ground water sustains annually can be no greater than that of the contribution to it annually. [Thus] [t]he draft for general irrigation would need to be at a rate considerably less than this small rate of replenishment, to avoid serious lowering of the water plane (to prevent early depletion), and such a very limited supply, even if obtained at no expense for lifting, would be of no material benefit in irrigation.

In short, Johnson concluded, "[a]gainst the High Plains,... the absolute verdict must be that they are nonirrigable" (ibid.).

Thirty years later, the noted historian of the plains, Walter Prescott Webb, concurred. Comparing the groundwater reserves of the plains to a bank account, Webb explained that to maintain a balance, withdrawals must match deposits. "If the withdrawals should be appreciably increased for purposes of irrigation, the bank account would decrease correspondingly ... and since the amount of deposit is fixed on an average, the loss could not be made good" (Webb, 1931: 331).

But neither of these experts could reckon with the determination of the farmers and government officials to remake the plains into an irrigated paradise. Charles Dana Wilber stated it succinctly: plains farmers believed in the "grand consent or, rather, concert of forces – the human energy or toil, the vital seed, and the polished raindrop that never fails to fall in answer to the imploring power or prayer of labor" (quoted in Smith, 1950: 182). For every Powell and Webb, there was an opposing voice asserting that, given insufficient rainfall, "irrigation waters could be produced in unfailing quantities" from underground supplies, and moreover, "that water is almost, or entirely, free from minerals that are injurious to plants, and the level to undulating topography and fertile friable soils are excellent for utilizing irrigation waters" (Browne, 1937: 169). Although the farmers had a very strong sense of obligation to future generations and to wise stewardship of the land and its resources, they also held an equally strong conviction that a way would be found. If the water was not inexhaustible, it could be made to seem that way with greater investment in pump technology and energy outlays.

The belief in the ability always to find a way, however unfounded, has taken on different expressions over the years: In the 1950s, limitless energy was key; in the 1960s and 1970s, high crop prices and extravagant water importation schemes were the answer; in the 1980s, world commodities markets and price supports kept the water affordable as depths to water increased; and finally in the 1990s, plans to tap deeper fossil water reservoirs, along with refined conservation strategies, seem to provide the answer to the question of where the water will come from.

The reality of groundwater reserves on the plains was accurately depicted by Johnson. The High Plains Aquifer system is a vast, irregular configuration with widely varying depths to water and varying saturated thicknesses. The great variability in depth to water across the High Plains Aquifer meant that reliable pumping technology was essential to maintaining a steady flow of water to the surface. Depths to water across the Llano alone ranged from a metre or so in the east to several hundred metres in the west. The application of motor-driven technology to pump design enabled farmers to reach depths to water inaccessible from windmill pumping. Although the technology was available around the turn of the century, the expense and difficulty of maintaining the equipment limited its use. Energy sources were problematic. Steam-driven pumps needed constant surveillance, and the alternative, electrically powered pumps, required electrification and power lines strung out to the fields (Bowden, 1977). These early models were cumbersome, prone to breakdown, and inefficient, making them very expensive to operate (Green, 1992).

The internal combustion engine, imported to the United States in the early years of the twentieth century, proved promising. At roughly US$4,000 (in contemporary dollars), irrigation plants were an investment that farmers wished to put off as long as possible (Bowden, 1977: 116). And, with the mostly good weather of the early decades, it was easy to postpone that investment. As the pumps were made more reliable and their range increased, it would become more feasible to invest in them. Nevertheless, there were only a few hundred "pumping plants" across the southern plains by the outbreak of World War I, drawing mostly from the so-called shallow water belt (Green, 1992). Interest in irrigation lagged through the 1920s, as higher than normal rainfall continued and crop prices dropped. With the drought of the 1930s, those who could afford to invest in irrigation did so, as boosters ceaselessly promoted its benefits. The loosened credit brought by government relief programmes made the investment affordable. With the drought, all crops needed extra water, not just those that could not be successfully dry-farmed. Soon reconditioned automobile engines were powering pumps at half the cost of the precursor models ten years earlier. When a simple, direct-drive way of linking the motor and the pump was devised, efficient, dependable, and relatively inexpensive pumping plants became feasible. By 1940, there were roughly 2,200 irrigation wells, watering over 100 hectares (about 250,000 acres), a huge increase over the 170 wells and perhaps one hectare total (under 3,000 acres) in 1930 (Bowden, 1977: 119).

Irrigation did not become truly widespread until after World War II. Not only were war-era innovations improving technology, fuel was also becoming cheaper. The discovery in the 1950s of extensive natural gas fields in the southern plains brought plentiful and inexpensive fuel right to the farmers' fields. Without the severe drought of the mid-1950s ("The Big Thirst"), though, it is doubtful that widespread and eager adoption of irrigation agriculture would have caught on quite so quickly.

All aspects of irrigation technology continued to evolve rapidly. The development of cheap and lightweight aluminium pipe after World War II meant farmers could single-handedly reconfigure their irrigation schemes quickly as conditions warranted; the new pipes also meant that less optimal land could be irrigated, since farmers no longer had to rely on gravity-flow ditch systems. Putting the pipes on wheels, an innovation introduced in 1950, made placement even easier (Green, 1992). Shortly after that, the centre pivot sprinkler was introduced. Irrigation expanded as farm credit programmes made the financing of irrigation systems possible and the belief spread, in the face of the facts, that the aquifer was a "vast underground river, continuously recharged with snowmelt from the Rockies"; farmers "exchanged stories about ... incidents of blind trout ... and other such evidence (found in) the irrigation ditches" (Urban 1992: 206).[1]

Irrigation and regulation on the Llano

Archeological evidence exists showing that irrigation has been practised on the plains for centuries, long before European arrival. Without any formalized consideration of whose water it was or who had the right to use it, irrigation continued on a small scale in the recent past, with windmill-driven pumps bringing water up from the shallow depths to water cattle, for example. Large-scale, widespread irrigation came about with the development of technology to gain access to non-artesian water and water at lower depths.

The move to regulate the use of groundwater coincided with the development of irrigation technology. In Texas, the two events are only marginally related. The move to institute some sort of control over groundwater use on the Texas Llano was fundamentally a move perceived to preserve the independence of the residents of the Llano against the intrusive action of the state, and potentially the federal, government (Brooks, 1996). The people of the Llano were acting to preserve an identity, steeped in a way of life.

The move to manage groundwater

The threat of recurrent drought throughout the first half of the twentieth century seemed to be eliminated by the access to groundwater in the years immediately after World War II. Technologically advanced irrigation systems and powerful new pumping plants provided the steady rainfall the region lacked. Throughout the 1950s, wide-scale withdrawals provided the water. The unchecked water withdrawals soon began to affect the aquifer, increasing the depth from which the water was pumped, and the costs. As early as the late 1940, concerns were being voiced about depletion, although the dominant view was still that a vast network of flowing rivers existed underground. But those concerns were enough to spur debate about the possibility of some sort of imposed management.

Larger state politics also were playing a role. Throughout the 1940s there had been several attempts by the Texas state legislature to declare underground waters the property of the state – surface water had been so designated earlier. As early as 1917, the Texas legislature had approved a so-called conservation amendment to the state's constitution, which required the conservation of "all of the natural resources of the state" (Cisneros, 1980). Although the amendment enabled the legislature to enact conservation laws, none were created with regard to groundwater. In 1937, in response to the expansion of irrigation in the state, the first attempt at a bill enacting comprehensive groundwater management was presented to the state legislature and failed. Three subsequent attempts in 1939, 1941, and 1947 also failed (Templer, 1992). Then, in the winter of 1947, legislation was raised to declare groundwater property of the state in the Winter Garden region only. The Winter Garden, an area comprising five counties in central south-west Texas, was a highly productive agricultural region, and the legislation was considered essential to preserving the water used in irrigation.

This move by the state legislature alarmed the people of the Llano. Perceived as a first step towards the eventual control of the High Plains Aquifer, the representatives of the Panhandle/Plains objected to the passage of the bill. The Winter Garden Bill was passed over their objections, and later in that legislative session, another bill was presented, the so-called Jameson Bill, which proposed making all underground water the property of the state. Met with outraged opposition, the Jameson Bill failed in committee. But the die had been cast, and the move toward some form of regulation begun.

Water management policy, and particularly groundwater management, in Texas had been marked by a generally laissez-faire attitude. This was expressed judicially in 1904, when the Texas Supreme Court found that landowners had an essentially unrestricted right to withdraw water found beneath their land (Cisneros, 1980).[2] In a policy characterized by legal scholar Johnson (1982) as "a policy of ... non-intervention," the state adhered to the common-law principle of ownership, which provided total ownership of all resources pertaining to a parcel of land. Operationally, this principle, characterized as "heaven high and hell deep," was followed by users of groundwater throughout the state, but was especially enthusiastically embraced on the Llano. The notion of rights as something to be conferred by a civil power did not sit comfortably with the water users of the Llano. Rights of use were instead earned by virtue of having wrested the land from the grip of Nature and transforming it into productive farmland: these were rights "earned by blood."[3]

Concern about state or even federal encroachment on water use was heightened nonetheless. Reasons to be apprehensive were found in the state's growing role in everyday life and the possibility of the federal government's involvement. Proposed state water projects, such as the damming of small rivers in the Texas Panhandle, were characterized by some as "creeping socialism" and admonitions were raised about becoming too dependent on state help.[4] In this climate, some on the plains thought it would be prudent to develop a home-grown strategy for dealing with what seemed to be the inevitable state intervention in water use.

The Llano representatives to the state legislature in Austin began to formulate a legislative response in 1948, in a move designed to pre-empt any further action by the state. Locally based management plans were drafted and submitted to the legislature for consideration. One of these proposals, House Bill 162, also known as the Holt Bill, was passed by the Texas Senate in April 1949. It was then sent to the Texas House who passed the bill in late May 1949. Specifically, the Holt Bill proposed codification of four basic principles: "To recognize individual ownership of underground water, to conserve, preserve, and protect underground water, to prevent waste of underground water, and lastly, to recharge underground water." The vehicle for this would be locally based groundwater management; the legislation provided for local elections creating a groundwater management district on the Llano. Farmers and property owners would vote in county-wide elections to decide whether to include themselves in a management district. The district or districts would be locally run, with board members and officials elected for a specific term of office, and residents of a given county voting for establishment of the district would voluntarily submit to regulation of their water use. The four basic points of the legislation were considered the foundation and thus the guidelines for any prospective management policies.

The late 1940s and early 1950s stand out as an era of great activity around the issue of establishing groundwater management. The beginning of the transition from a totally individualistic, self-interested, unregulated resource scenario to a centrally based, voluntaristic system of groundwater management was signalled by the concern raised over state actions to control the water. Widespread debate over the appropriate role of authority in an individual's exercise of his or her rights ensued, a debate whose subtext was the importance of maintaining a particular cultural identity honed by a generation of pioneers (Brooks, 1996).

Because so very much was at stake in the decision to create a groundwater management district on the Llano, the months leading up to the elections became a time for considerable questioning and inquiry. Compliance with programmes requiring changes in farming practices had a mixed history in the region. The farm relief efforts of a decade earlier were successful over the long term if farmers believed they were still able to opt for their best judgements, it can be argued. The soil conservation practices introduced by government experts were less readily adopted and were quickly abandoned once they were no longer perceived to be necessary. For farmers to adopt the water conservation measures implicit in creating a management district would mean accepting limits on their practices and on their closely held beliefs.

The debate and the issues

The debate took place over a period of several months, and was held largely through open meetings and hearings sponsored by the Texas Board of Water Engineers. These hearings were well attended and transcripts were reprinted in the local newspapers. Farmers and other property owners needed little encouragement to voice their opinions both for and against the district, and experts and local officials were there to answer questions. The meetings were generally held at local schools, community rooms, or town halls. The participants were relatively well-off: property ownership was a condition of electoral participation and the standard of living was fairly high on the southern plains.[5]

The issues that mattered most to the people of the Llano with regard to creating a groundwater management district make sense in light of their perception of their own history. Many of these "hardy sons of pioneers" were the first generation of Euro-Americans to have been born on the Llano. Nearly everyone was no more than one generation removed from the original settlers. The image of the lone pioneer – with his family of course – battling the fierce climate and, after persevering through repeated hardships, triumphant amidst great material wealth, cast a long shadow over the debate.

Consequently, the farmers of the Llano were most concerned about the erosion of their control over their own affairs. The locus of the power inherent in the proposed management district was critical – state-level or, worse yet, federal-level agencies were anathema on the Llano. Also important was the notion of being free to perpetuate certain farming and grazing practices and other customs, all of which contributed to a generalized concern of maintaining a sense of identity historically derived.

Whether or not the district would make a difference in the practical life of the aquifer was also important. "Water ... [is] the life blood of the plains," banner headlines proclaimed, in support of the district's formation.[6] Because of that widely acknowledged truism, responsible action was required of the people of the Llano in order to preserve a way of life. That argument was countered by those who maintained that "the fact is, the land owner himself does conserve and protect his property, both his water and his land."[7]

Ultimately, 12 of the 21 counties participating in the election voted to create a groundwater management district, roughly 1.9 million hectares (4.8 million acres) in size. (The decision was binding only for those counties that voted in the majority for the district.) All other counties on the Llano were not included. Many aspects of the district needed to be implemented before it could begin to act: boards of directors and committee people had to be elected, by-laws written, and general district organization decided. By January 1952, elections were held to appoint the leaders and committees of the district, and a new era on the Llano had begun.

The First Underground Water Conservation District

High Plains Underground Water Conservation District No. 1 (HP#1 hereinafter) was the first formalized recognition of the transient nature of resource stability on the Llano. Although compliance was essentially voluntary, and wells incapable of pumping a minimum of 378,500 litres (100,000 gallons) per day were exempted, HP#1 had certain powers that could, on paper, impose real limits on groundwater withdrawals within the designated "underground water reservoir," the boundaries of which were coterminous with the district. The order issued by the state of Texas creating the district detailed its powers and responsibilities:

1. Formulation, promulgation and enforcement of rules and regulations to conserve, preserve, protect and recharge the underground water in the designated reservoir;
2. Formulation, promulgation and enforcement of rules and regulations to prevent waste of underground water;
3. Requiring of permits for drilling and completion of wells, and the issuance of said permits when the conditions are in compliance with the requirements of the above;
4. Provide for well spacing, to prevent excessive drawdown or reduction of artesian pressure. However, no one will be denied permission to drill a well on his or her own land, subject to numbers 1 and 2;
5. Requiring records to be kept and reported on drilling and completion of wells, and also drillers' logs;
6. Acquisition of lands for water projects (dams, drainage, recharge projects) but never of water, underground or surface;
7. The conduct of underground water surveys by professional water engineers;
8. Development of comprehensive plans for most efficient use of underground water and to prevent waste of that water, and to carry out research, collect and maintain data, and publish that information for the benefit of the water users and other interested parties;
9. Enforcement of rules and regulations, through the courts, by injunction or other remedy, after notice is duly given.

(Text of Order Upon Hearing and Granting Petition To Create High Plains Underground Water Conservation District #1, 9 August 1951).

The authority of HP#1 was challenged by many disgruntled farmers, particularly on the issue of tax assessment – the district was authorized to levy a tax to fund the district's activities. But, overall, because in reality the district's powers were quite limited, support for it grew. Over time, elections were held for other counties to decide on whether or not to join. (In 1999, the district extended to 15 counties on the Llano.) The impact the district's activities had on water use practices was always constrained by the affirmation of individual private ownership of groundwater by the owner of the land, although the district reported a decrease in water application rates to 3,049 cubic metres per hectare per year (one acre-foot per acre per year) in the 1980s from 6,098 cubic metres per hectare per year (two acre-feet per acre per year) in the 1950s (Templer, 1985).

One area in which HP#1 has had a major impact is in the fulfilment of its duty to collect and make available data on groundwater. Its monthly newspaper, *The Cross Section*, has a circulation of around 6,000 readers and serves as a forum for the dissemination of information about innovations in irrigation techniques, crop research, and conservation practices. In addition, the district distributes annual maps detailing water use and decline over the district's territory. Water use might have been more sensitive to market demands than conservation practices, but at least the farmers knew exactly what they were doing to the aquifer.

Part of the impetus behind keeping the water users and others so well informed was the belief that "the surest way to conserve water is to make each farmer and land owner anxious to police himself" (May, 1962). Throughout the 1950s, 1960s, and 1970s, water use expanded rapidly, however. Clearly, farmers were responding to the high prices being paid for crops, the availability of credit to finance expensive irrigation works, inexpensive energy to run the pumps, and an almost overwhelming sense of destiny.

The role of water institutions throughout the 1950s and early 1960s centred around the promotion of conservation, support for research into aquifer recharge, and public education about the importance of water to the region. This was an era of great optimism in the United States: wealth was increasing and was reaching more and more people (Potter, 1954). On the Llano, people were experiencing economic growth as well. In 1955, for example, Lubbock, the "Hub City of the

South Plains," ranked first in the nation among cities of comparable size (around 120,000 population) in per capita income, per family income, and effective buying power (Lubbock Chamber of Commerce, 1955). Belief in technological innovation was strong as availability of consumer goods and industrial output soared.

When alarms were raised over the dropping water levels – as much as three metres (nearly ten feet) in some parts of the Llano in 1954 alone – the district moved to educate farmers about federal water importation schemes.[8] The appropriate response would take full advantage of technology and capital. These massive projects relied on essentially unlimited energy inputs and perceived water surpluses and involved hundreds of miles of canals to bring excess water to areas of water deficit. In particular, HP#1 strongly advocated the Texas Water Plan, formulated and eventually submitted to electoral decision in 1969. The Plan proposed a series of canals to deliver roughly 11 billion cubic metres (about 9 million acre-feet) to west Texas and eastern New Mexico from the Mississippi River system, a distance of over 1,200 kilometres (about 800 miles) with a lift of more than 1,000 metres (3,000 feet). The energy budget for the plan was estimated at roughly 40 per cent of Texas' electricity usage in 1970 – 7 million kilowatts (Bowden, 1977: 121). The imported water, it was planned, would be stored in the dewatered aquifer, replacing the exhausted groundwater with the inexhaustible Mississippi, assuming its riparian states would allow it. Promoted as the answer to any water needs in the future, the plan was nevertheless voted down, probably due to its 9-billion-dollar price tag and the spectre of federal involvement (Urban, 1992).

The district's role

Despite its activism on behalf of the failed importation scheme, the district regrouped and refocused its efforts, after the proposal's defeat, on conservation of groundwater to offset the annual drop of roughly a metre (three feet) in the water table.[9] Since technological answers on a massive scale were clearly no longer an option, attention was turned to the small scale. Continued innovation in irrigation technology seemed to provide a means of using the water more efficiently, and the district was committed to sponsoring research in irrigation equipment and practices. Test plots were set up and demonstrations conducted. HP#1 representatives canvassed the area and pointed out waste problems, like open and unlined ditches, furrow irrigation, and unreturned tailwater (runoff from irrigated fields). The district also used its enforcement powers to issue citations to farmers who neglected to correct wasteful practices.

Nevertheless, water use expanded in the 1970s, with wildly increasing withdrawals; HP#1 also grew in prominence. Although the Board of Directors argued that the district was an active force in groundwater management, its role was still chiefly one of education and research.[10] Significant advances in the efficiency of irrigation using new techniques and more refined equipment, particularly the increased spread of centre pivot irrigation systems, contributed somewhat to conservation of water; still, the 1970s was an era of unprecedented withdrawals from the aquifer. With the onset of the energy crisis in the late seventies, however, and the shifting world market for the agricultural products of the Llano, farmers could no longer afford to pump their water from such great depths in such great quantities.

Water withdrawals continued to decline throughout the 1980s, and the agricultural economy adjusted as well. The district's role as educator grew more important, with programmes on water conservation designed for the public school system, expanded demonstration projects, and more research into increasingly efficient irrigation systems. An example of these is a modified form of centre pivot irrigation, the Low Energy Precision Application System, or LEPA, which reduces water waste from 40 per cent with conventional furrow irrigation to about 5 per cent (HPUWCD #1, 1995). However, HP#1's ability to restrict irrigation or curtail certain inefficient or wasteful practices remained quite limited.

Water use and exploitation

It could be argued that irrigation expanded in advance of the data necessary for its efficient or even proper use. Water use was unconsidered and wasteful. Crops requiring no more than 30 to 45 centimetres (12 to 18 inches) of supplemental water were routinely being irrigated with over 60 centimetres (24 inches) of (non-renewable) groundwater: clearly if a little extra water was helpful, then a lot more would be even better (Urban, 1992: 206).

Groundwater exploitation is perceived to have hit its zenith in the 1960s. There were 71,000 wells, irrigating about 2.6 million hectares (6.4 million acres). Agricultural production surged in the region: 34 per cent of Texas cropland was plains farmland, and 69 per cent of all irrigated cropland in the state was on the plains (Urban, 1992: 207). Not surprisingly, withdrawals from the Ogallala peaked in 1977 at roughly 10.16 billion cubic metres (about 35,893,440,000 cubic feet or 8.24 million acre-feet), according to Urban (1992).

Since the 1970s, groundwater use has steadily, and sometimes precipitously, declined. Some areas of the Llano have reverted to dryland farming practices, ending irrigation altogether. There are many reasons behind this transformation: energy costs, changing demand, increasing depths to water, and the institution of formalized management. In sum, the inexhaustible water supply was giving out, for all practical purposes.

Conservation practices in the post-war years

Many of the conservation practices brought to the Llano through the relief programmes of the 1930s were abandoned with the return of adequate rainfall in the 1940s, and the burgeoning demands for grain and fibre resulting from the United States' war effort and the needs of Europe. The Shelterbelt programme was unsuccessful on the Llano chiefly because the rainfall was insufficient to support lush tree growth. Conservation tilling practices were neglected as farmers were once again exhorted to plant fence line to fence line to meet wartime and post-war demand.

The Dust Bowl consequences were considered lessons learned; the general consensus was that times had changed and things were different now. One major difference was the expansion of irrigation fed by the exploitation of groundwater. It was difficult to persuade farmers to keep fields in grasses when the prices for irrigated cotton and cattlefeed were soaring.

Some of the lands that had been retired and entered into Department of Agriculture reserve programmes through the Land Utilization Project were allotted to the Soil Conservation Service. These lands, some 4.4 million hectares (about 11 million acres) across 45 states, included some of the most severely damaged in the Dust Bowl region (West, 1990). The Soil Conservation Service projects included reclaiming these lands for improved range and grazing, recreation, wildlife habitats, and watershed protection, employing farmers and others on relief.

The ultimate goal for these Land Utilization Projects, however, was not permanent retirement but rehabilitation out of farmland and into rangeland (West, 1990). Eventually these grasslands were folded into the National Forest System. Problematic at best, the incorporation of grasslands under forest management policies has led to a confused mission regarding the preservation of these lands. In the end, though, a major aim of the grasslands programme was to promote grasslands agriculture.

The most active conservation programmes on the Llano today concern irrigation practices. Innovations in water application and capture of return flows from fields dominate. Perhaps the most successful land conservation programme has been the passive retirement of farmland due to bankruptcy, lack of water, and abandonment. What happens to that land is of grave importance to the future of the Llano, because it may be showing us what lies ahead for the region.

Institutionalized assistance

Given that the water supply would always be problematic, whether through feasibility of pumping lifts, market demand for crops, or any of the other unpredictable impacts on farming in essentially untenable environments, government farm subsidy and support programmes have become an essential structural support system.[11]

An entire superstructure of farm programmes had their beginnings in drought relief efforts of the 1930s, largely owing to the gathering strength of the organized farming groups, slowly evolving into a powerful farm lobby (Hansen, 1991). Price support programmes became a fixture of agricultural life, protecting the farmers against fluctuations in markets at home and abroad. Although production controls, federally imposed limits on planting, seemed to address some of the underlying causes of the farm crisis of the 1930s, in reality they treated only the symptoms (Hansen, 1991). The onset of World War II and the concomitant soaring demands for grains and fibre, and the return of the rains that enabled plains farmers to meet those demands, had turned the farm crisis around. Despite the upturn, the farm lobby managed to keep price support and parity pricing programmes active.

The booming post-war economy had wide-ranging impacts on agriculture, accelerating its growth as well. This time, when the "Big Thirst" returned in the mid-1950s, threatening farmers anew, the potential disaster wrought by overproduction and overutilization of marginal land was mitigated substantially by loosened credit, crop price supports, and irrigation. More than half of the total land irrigated on the Llano by 1955 had been brought into production since

1950 (USDA, 1955). Irrigation had changed the impacts of the drought, thanks in large part to federal government programmes encouraging its development. The Farmers Home Administration was responsible for thousands of direct loans to farmers to assist them in installing irrigation works. These loans could be used to cover drilling costs, for equipment, for levelling lands, and to construct ditches and canals (USDA, 1955).

The credit–capitalization–investment spiral has been maintained to the present. What's more, the underlying investment in producing a bushel of wheat has increased probably a hundredfold, but that bushel of wheat still commands roughly the same range of prices it did in the 1950s.[12] Institutionalized farm credit has become, of necessity, an essential, central fact of farm life. What remains to be seen is the impact of federal revamping of major farm programmes in the 1990s. With many farm price support programmes due to be phased out by the turn of the twenty-first century under the 1996 Freedom to Farm Act, one can only speculate on the consequences for regions like the Llano, whose uncertain water supply has been supplemented by certain government intervention to maintain the viability of the agricultural economy. Federal institutions supporting farm production on the Llano quickly became a welcome and necessary substitute for adequate and reliable rainfall. What will play that role in the future is unknown.

Intensification of cotton agriculture on the Llano Estacado

With confidence in the new-found water source, and the belief that the minimal management system that had been erected would ensure the wise and enduring use of that water, farmers on the Llano approached agriculture with renewed energy. During the early part of the century, cotton culture development had been a process of "adaptation to the prevailing natural conditions" (Gordon, 1961: 108). The research conducted by the State Agricultural Extension Service and individual farmers consisted mostly of breeding cotton varieties that thrived in the particular soil and climatic regime of the region. In the post–World War II environment of innovation and progress, a period of intensification and state involvement began (Cochrane, 1979). Land productivity was increased through the development of biochemical technology and labour productivity was increased through the substitution of capital, in the form of ever more efficient mechanical farm equipment, for labour (Holland and Carvalho, 1985). Both avenues led to increased dependency upon non-renewable resources and an elaborate institutional regime of credit and technology development.

During and after World War II, rising domestic and foreign demand for most commodities, including cotton, spurred an increase in irrigation and cotton production. The new pumping technology made access to the deep water reserves (down to 90 metres, or 300 feet) of the Ogallala possible, tapping what appeared at the time to be that nearly endless water source. With ample groundwater and a plentiful supply of natural gas from the Texas gas fields, irrigation wells increased at the rate of approximately 2,000 a year across the Southern Plains area (Gordon, 1961).

Mechanical harvesters brought another important technological change in cotton culture on the Llano Estacado. Their use increased with the labour shortage created by World War II. By 1951, approximately 21,000 of the machines were in operation, enough to harvest all the cotton grown (Jones, 1954). In addition, improvements in ginning machinery made machine-harvested cotton more feasible. Mechanical stripping created problems of trash in the cotton as burrs, leaves, stems, and other material were gathered with the bolls by the machine. Storing the cotton on the field prior to ginning exacerbated the problem. In 1950, a local person, Ennis E. Moss of Lubbock, invented the lint cleaner, and in 1958 he began manufacturing machines to clean the lint after ginning.

The Korean War again increased demand for American cotton and existing surpluses were quickly depleted. High market prices encouraged more planting and, consequently, irrigation expansion. Surpluses were quickly built up and the Secretary of Agriculture announced marketing quotas for the 1954 crop. (Marketing quotas were in effect until phased out during the 1990s.) Planting allotments (which were also eliminated by 1990s farm policy) were instituted with the marketing quotas and were particularly restrictive where the most recent development had occurred. Farmers responded by increasing fertilizer and other inputs, to increase yields on the allowed area. As a result, total yields remained nearly the same even though acreage declined. Parity prices virtually guaranteed commercial farmers on the Llano Estacado favourable commodity prices for their crops, and a substantial profit. With increasing yields on less land, and essentially guaranteed crop prices, a very productive and profitable agricultural industry developed. Lubbock emerged as a major cotton market and centre for cotton and cottonseed processing. Many gins and pressing mills were built in the area and Lubbock became the "cotton seed oil capital of the world" (Hill, 1986). Agricultural implement and equipment suppliers and manufacturers were booming, as were cattle-feed suppliers and meatpackers. Fuelling the boom, manufacture and installation of irrigation equipment also became big business. The 1950s and 1960s emerged as the heyday for agricultural producers and their supporting industries.

The federal government and farm policies

A major factor in maintaining the farming economy of the Llano was the involvement of the federal government in financial support and planting area controls, dating from the first version of the Agricultural Adjustment Act in 1933. Through the early 1970s, cotton farmers were guaranteed a good income from the USDA programmes if the market or the weather were to fail them. Since the mid 1970s, though, the US has moved towards a deregulated and unrestricted cotton industry. The cotton provisions of the 1973 Agriculture and Consumer Protection Act were drafted to have little or no effect on the level and stability of farm prices, consumer prices and supplies, or world markets. Cotton disaster payments were about the only provision of the Act to bolster net farm income in 1974 and 1975. At the same time, farming costs, and particularly irrigation costs, were increasing.

As groundwater levels fell, more energy was required to lift water to the surface and specific yields declined. By the early 1980s, energy costs had increased by an order of magnitude from the late 1960s. Frequently, irrigation costs constituted more than half of the variable costs of crop production on the South Plains of Texas (HPUWCD #1, 1983). According to Hill (1986), the agricultural economy of the area was sustained chiefly by the general inflation experienced in the United States from 1973 through 1981. From 1980 to 1986, the farm value of cotton was not enough to cover its production costs. Without the help of government payments, cotton producers would not have earned any profits after paying all costs, including returns to land and unpaid family labour. The solvency of cotton farmers as a group increased slightly toward the end of the 1980s, most likely owing to government support programmes. On the other hand, a sizeable subset of cotton farmers faced negative net incomes as income and cash flow deteriorated (Chen and Anderson, 1990). But surely times had changed and the exuberance of production and price levels experienced in the 1950s, 1960s, and early 1970s was gone.

The intensification of the feedgrain–cattle complex

Cattle producers suffered along with farmers and tenants during the disastrous 1930s. Forced retirement of crop land and conservation projects which promoted reseeding and range restoration were thought to benefit ranching on the Llano, but reinvesting in cattle was seen as risky, given the recent history. Overgrazing had played its part in the environmental devastation of the region, and cattlemen were cautious about taking big losses again. Irrigation expansion had its impacts in this sphere as well. With steady irrigation, feedgrain production became a feasible way to support grazing, and set in motion the development of the feedlot industry on the Llano and the Northern High Plains of Texas. It was not yet apparent that the increased mechanical and technological intensity of agriculture that occurred after World War II would also push many farmers to accrue considerable debt for tractors, improved irrigation equipment, fertilizers, and agrochemicals (Green, 1973).

Cattle spend most of their lives grazing on pastures of native grasses, forage crops, crop residue (often corn), and wheat. Wheat is an integral part of cattle production, not simply because feed contains wheat, but because cattle graze on protein-rich emergent winter wheat in the fields. Since at least the 1940s, it has been common practice for Texas cattle to "run on wheat." Winter wheat is planted in September or October and cattle are put onto the fields during winter months to graze, or "run on," the growing wheat. As long as the cattle are removed by mid-March, little or no damage is done to the wheat, and normal yields can be expected when the crops are harvested in late June or early July. Most harvested wheat on the Llano is sold for food processing (bread and pasta), not cattle feed. Cattle graze on crop residue as well, particularly corn, during winter months. Dryland hay is cut only once and stored for winter feeding, but late summer regrowth may be sufficient to augment cattle diets during the winter. Cattle producers may also feed cattle some feedgrain as well, but it is not until cattle are moved to feedlots for fattening that they start consuming substantial amounts of feedgrain.

Cattle on feedlots are fed rations high in corn and wheat, and to a lesser extent, sorghum, alfalfa, and corn silage (Beckett and Oeltjen, 1993). Nationally, corn is the preferred fattening grain, followed by wheat, but significant regional variations exist. In Texas, for example, approximately 17 per cent of the ration is sorghum (Beckett and Oeltjen, 1993: 823). Wheat can be substituted for corn or sorghum when corn prices rise. Cattle gain hundreds of kilos in the five to seven months they are fed in the yards.

As fed-cattle operations and meatpacking plants moved to the High Plains, the geography of doing business changed for Texas cattle producers. Before this shift, Texas cattle producers typically shipped their cattle by rail to California or Arizona, and to a lesser extent to eastern Texas (Fort Worth) and Kansas City, for fattening and sale at auction to meatpackers. The irony (and lost economic opportunity) of shipping both Texas grain sorghum for cattle fattening and Texas cattle for fattening and slaughter hundreds of miles away was recognized by many in the cattle industry (Ball, 1992; Nall, 1982). The irrigation boom of the 1950s, increasing the production of feedgrains, helped shift the industry's locus.

Thanks to cheap and abundant groundwater for irrigation, feedgrains remained abundant and relatively inexpensive in the 1960s at the same time as consumer preference was shifting to grain-fed beef. In 1970, grain-fed beef accounted for two-thirds of all beef consumed in the United States (Stephens, 1982). Irrigated sorghum acreages remained high through the 1960s, but fell dramatically in the 1970s. Over 800,000 irrigated hectares (2 million acres) of sorghum were grown in Texas during the early 1960s, but that figure was more than halved by 1978. On the Llano, farmers irrigated more than 526,000 hectares (1.3 million acres) in 1964, but less than 162,000 hectares (400,000 acres) in 1978 (see figure 5.1). Corn was replacing sorghum as the desired cattle feed. A highly profitable crop, as mentioned earlier, corn became an important input to many processed foods, particularly as a sweetener and starch. Corn syrup and dextrose became increasingly popular as sugar substitutes. Corn was also exported both for cattle feed and for human consumption.

Despite government support and increasing consumer demand for beef, many farmers still faced cost-price squeezes during the 1960s (Green, 1973). Inexpensive grains during the 1960s permitted cattle producers to respond to increasing consumer demand for beef. Low grain prices, though, hurt wheat and cotton producers, who faced oversupplies due in part to European and Asian post-war agricultural recovery.

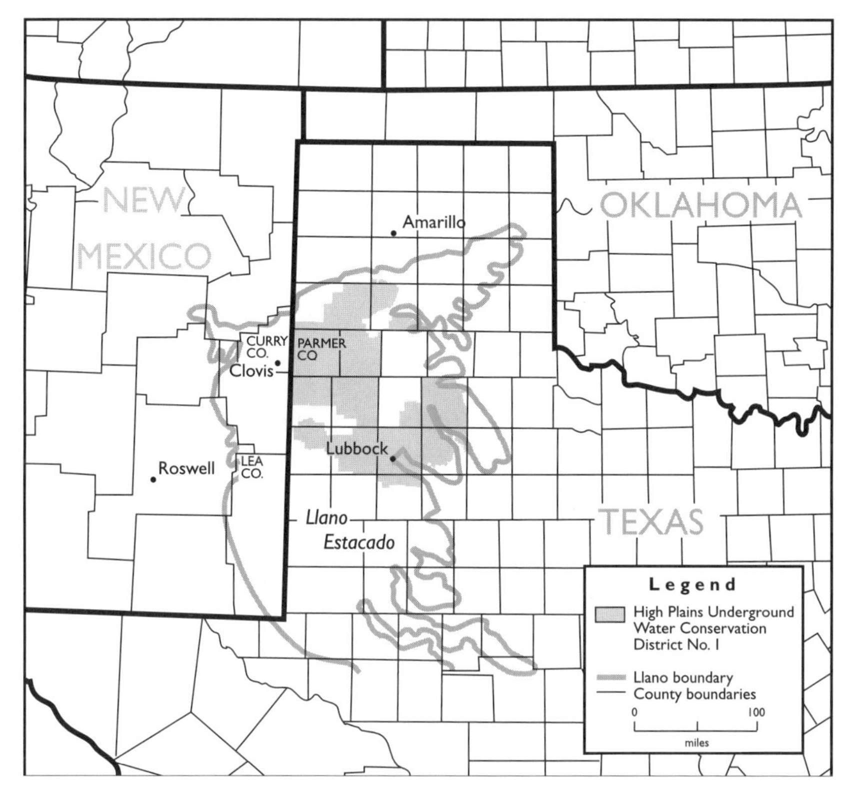

Figure 5.1 **Irrigated acreage: Study area (1964–1994)**
Source: US Agricultural Census.

In addition to low grain prices and farm debt, many farmers faced the first signs of groundwater depletion. By 1969, there had been over twenty years of intensive irrigation on the High Plains of Texas. More than 5.6 billion cubic metres (4.6 million acre-feet) of groundwater were consumed on the Llano in 1969 alone, actually a considerable reduction from 7.95 billion cubic metres (6.45 million acre-feet) five years earlier (TWDB, 1996b). Much of this irrigation was occurring in counties that would, as of 1980, report an increase in average depth to groundwater greater than 30 metres (100 feet) (Dugan et al., 1994). Cotton, the largest single irrigated crop, accounted for a considerable portion, but this was equalled by the combined total of sorghum, corn, and wheat (see figure 5.2). Feedgrains cultivation brought the cattle industry to the area, but was also depleting the groundwater the Texas fed-cattle industry depended on.

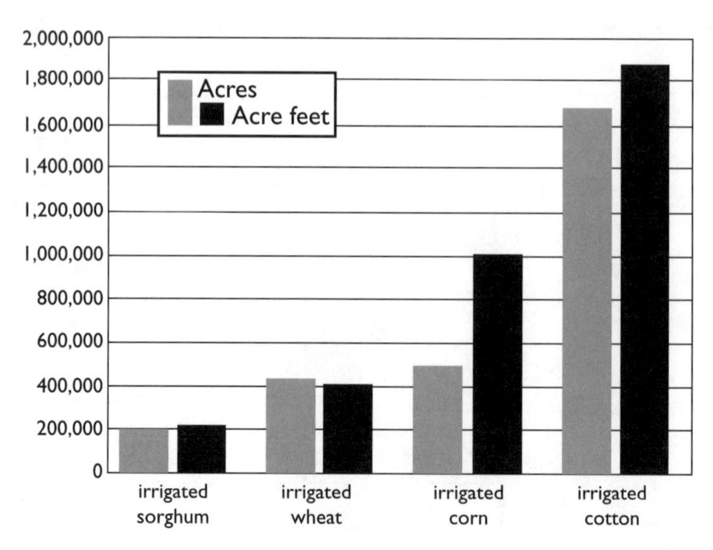

Figure 5.2 **Irrigated crop comparison: Study area totals, 1994**
Source: US Agricultural Census.

Green (1973) asserts that farmers responded to these conditions by increasing dryland crops and increasing livestock production. Irrigation data and cattle inventories indicate that there was a shift in irrigated crops (not a decline) and a considerable increase in cattle numbers on the Llano (see figure 5.3). Irrigated wheat acreage declined somewhat and irrigated sorghum acreage fell substantially after 1969, but corn made up the difference. Very little corn was irrigated prior to 1970 (less than 40,470 hectares or 100,000 acres), but nearly 162,000 hectares (400,000 acres) of corn were irrigated in 1974. Cattle inventories more than doubled between 1964 and 1969, in part reflecting the growing numbers of feedlots, but also indicating that farmers were responding to cheap grains and rising cattle prices.

In the early 1970s, cattle prices were climbing and the fortunes of cattle producers looked bright (see figure 5.4). Extremely large feedlots sprung up on the High Plains and meatpackers soon followed. Farmers modernized their equipment, increased chemical and synthetic inputs, specialized their production, and extracted tremendous amounts of Ogallala groundwater to increase yields of feedgrains, wheat, and forage crops. But modern intensive agriculture on the High Plains was not immune to international political and economic shocks, competition, or consumer preferences.

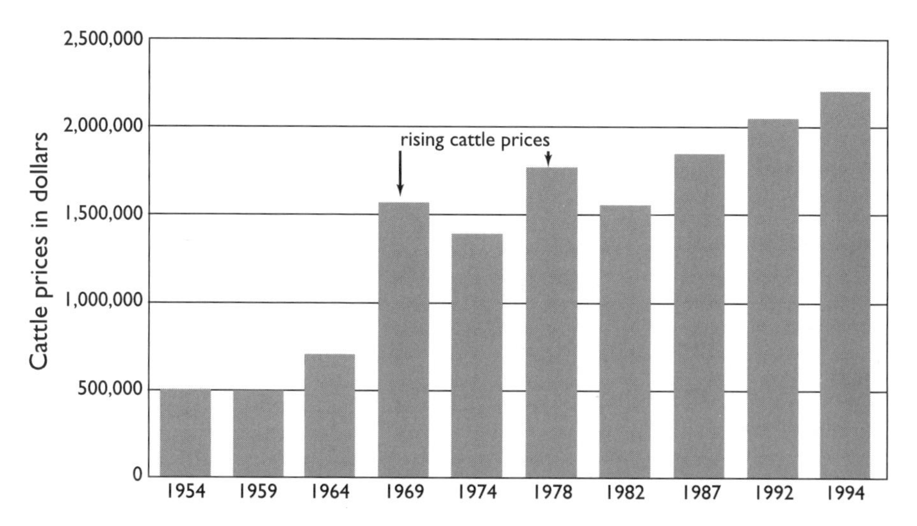

Figure 5.3 **Cattle inventory: Study area (1954–1993)**
Source: US Agricultural Census.

In 1973, beef prices were at an all-time high when "energy-food-currency shocks" (McMichael, 1992) rocked the US economy. Energy prices, and therefore groundwater pumping costs, jumped overnight. Irate consumers organized meat boycotts and President Nixon responded by freezing beef prices. The Soviet Union had just endured two disappointing harvests and a major grain deal had been struck in 1972 with the Nixon administration. As grain supplies dwindled through increased exports and decreased production, and prices climbed, grain-fed cattle producers faced rising costs. Many responded by liquidating their cattle stock in order to avoid paying for grain to fatten them. Recreating conditions of barely a century earlier, this rush to the market sent cattle prices tumbling. Average monthly prices for Amarillo Feeder steers (320–365 kilos or 700–800 pounds) fell from US$58.73 per hundredweight in August, 1973 (an all-time high) to below $30.00 in October, 1974, a mere 14 months later (Davis et al., 1994: 63).[13] Cattle inventories on High Plains feedlots also plummeted by almost half, from 1,686,000 in 1974 to only 909,000 in 1975 (Nall, 1982). This rapid and large-scale liquidation forced many feedlots out of business; the number of feedlots in north-west Texas with a capacity of 1,000 head or more dropped from 118 in 1973 to only 89 in 1977 (ibid.).

This cyclical pattern of cattle price increases followed by a rapid decrease is well known in the industry. A "cattle cycle," thought to

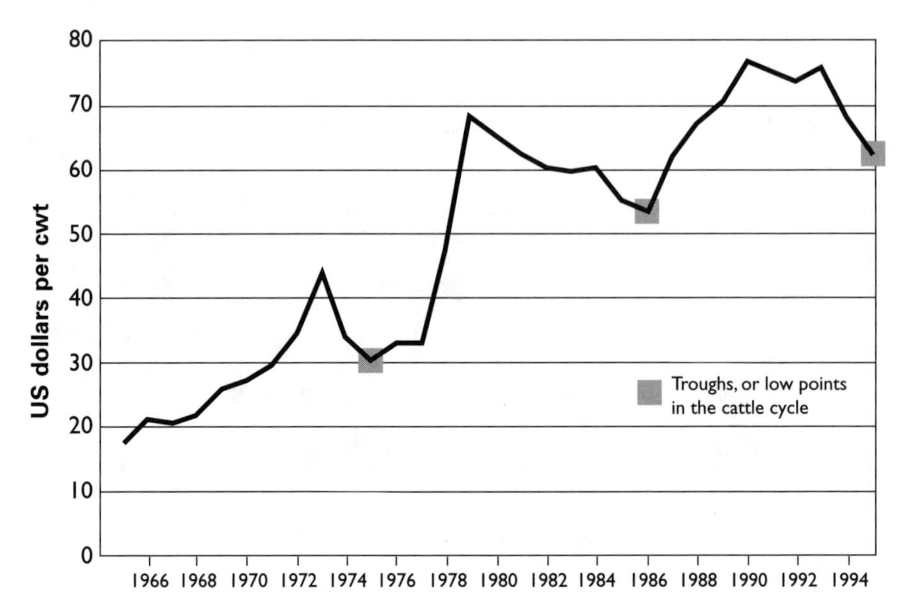

Figure 5.4 **Prices per hundredweight received for Texas beef cattle (1965–1995)**
Source: US Agricultural Census.

run 10 to 12 years, is the period from one low point in cattle prices to the next. Once cattle prices start to fall, farmers liquidate their stock to compensate for falling income or to avoid paying more to fatten an animal than the additional weight would bring at market. This liquidation leads to lower prices until farmers hold back their cattle to rebuild stock, leading to a gradual increase in prices (Sanderson, 1986). Of course, supply and demand are shaped by numerous complicated and interrelated actors and political economic forces, including grain prices, droughts, health concerns, and competition from other meats (*Successful Farming*, 1995). During the mid 1950s, a low point occurred after beef prices had jumped following World War II. The federal government lifted price controls and prices climbed from $15.00 per hundredweight in 1946 to $30.60 in 1951. But the drought in the 1950s created a scarcity of forage and farmers liquidated stock, causing prices to fall to $14.10.

Things were changing on the Llano. Water was getting more expensive and less reliable. Markets were shifting internationally and domestically. The ways of growing crops, particularly cotton, and raising cattle had to go in new directions as well to keep up. The resources that had been integral to success on the Llano in the 1950s and 1960s were undergoing transformations that echoed the evolution of the new environment on the Llano half a century before.

Notes

1. This echoed a popular sentiment earlier in the century that there was in fact an "Underground River," "one of the largest systems in the United States ... [which] starts in the melting snows of the Rockies, sinks below the surface and at the urge of gravity, starts southeast ... For hundreds of years this water has been flowing under the plains on its way to the Gulf [of Mexico] and mankind knew it not ... Now that this subterranean pipeline is being tapped, the plainsmen ... claim that they have the nearest ideal system of agriculture on earth" (Zenas E. Black, 1914, cited in Bowden, 1977: 120).
2. One reason cited by the Court was the very mysterious nature of groundwater, so "occult that its regulation was impossible" (Cisneros, 1980).
3. Statement by T.L. Wright, a Plains farmer testifying at a hearing regarding the implementation of a water management district on the Llano.
4. A series of editorials in the *Floyd County Hesperian* (16 January 1947) entitled "Ought to Have Dams" and "Not Too Big Dams" raised the spectre of the State Board of Water Control eventually "tak[ing] over the farmer's little irrigation outfits and tell[ing] him what he can and can't do after it is too late."
5. "Texas Farmer Is Leading Nation in Home Comforts," *Ralls Banner*, 4 January 1952.
6. Full-page newspaper advertisement sponsored by district supporters. The text of the advertisement was led by "People of the High Plains of Texas know that the water table is falling."
7. Text of testimony at a hearing about the district's formation, reported in the *Amarillo Daily News*, 29 September 1951.
8. The key role HP#1 would play in the administration of the water imported to the Llano was a prominent argument raised in the election of January 1967, which was called to allow three more counties the option of joining the district. A counter-argument was raised that if the water should actually ever be imported to the Llano, membership in the district would certainly not be the deciding factor in who received the benefit of the water. At an estimated cost (in 1977 dollars) of $430 to $569 per 1,234 cubic metres (one acre-foot), this was probably true, as who would be able to afford the water would be the salient factor (Urban, 1992).
9. Some areas have reported a drop of over three metres (10 feet, actually) in water levels at different times over the past 40 years.
10. In a scathing letter to the Executive Director of the Texas Water Development Board from the manager of HP#1 Frank Rayner, and signed by the Board of Directors, this point was made emphatically. "[I]t is obvious that the High Plains "leaders" have already taken the "institutional" initiative," Mr Rayner states (letter dated 4 March 1974).
11. The current status of many farm programmes is uncertain. Sweeping changes were enacted in the 1995–96 US legislative session; the impacts have yet to be realized. Price support programmes have been eliminated and crop subsidies,

so-called deficiency payments, are being phased out over a period of seven years.

12. The futures market was projecting a price of $2.50–$2.60/bushel for wheat in January 2000. Prices projected for wheat in January of 1955 were about $2.05/bushel. These figures are futures-based and not adjusted for inflation; nevertheless a comparison of inputs in unadjusted figures helps tell the story. Energy prices in late 1999 to early 2000 average around $1.40/gallon for gasoline, a basic input for running farm machinery, compared to roughly $0.25/gallon in the 1950s. (In the United States, one bushel equals 35.3 (American dry measure) and one gallon equals 3.79 litres.)
13. In the United States, one hundredweight equals 100 pounds (about 45.5 kg).

6

The transformation of the Llano in the post-expansion era: 1980 to the present

Ominous contraction followed the post-war years of unanticipated growth and expansion on the Llano. The limits of the groundwater supply were first ignored and then undeniably obvious, as depths to water increased rapidly over the years of the late 1970s and 1980s. The impacts of this were felt by cotton growers and cattle producers, and by the rest of the Llano's residents as well. Shifts in strategies and changes in practices, prompted by weather and markets, have marked the recent decades on the Llano. The future has also changed because of the reality of the water; what the Euro-American settlers imagined and the successes of the post-war era portended has been tempered by that reality.

Cotton on the post-expansion Llano

Today, cotton production pumps an estimated US$2.5 billion into the regional economy of the Llano, according to Eduardo Segarra, an agricultural economist at Texas Tech University in Lubbock. Llano farmers grow 60 per cent of the cotton produced in Texas, 15 per cent of the cotton grown in the United States, and 3.5 per cent of the world cotton crop. This gives the region the largest concentration of cotton production in the world (Daniel, 1995a: 3). Given that irrigation of the crop places one of the biggest demands on the aquifer system, cotton production has been a major driving force of environmental degradation or resource depletion. Globally and historically embedded and locally adapted, the production of cotton is a primary force in shaping the region's current economic and environmental status. The global market for cotton affords both significant opportunities and constraints for local producers dependent upon nonrenewable groundwater.

The global cotton market

Cotton has been grown on the Llano since the beginning of the twentieth century. Through the 1920s, cotton was a dryland crop (in the 1930s virtually all crops grown on the Llano were dryland farmed by default). After World War II, cotton started to emerge as a major cash crop, thanks to the sustained high yields made possible by irrigation. The overuse of water throughout that period of expansion led to the restrictions of the 1970s and 1980s. But the developments of technology and enterprise that occurred more than a hundred years ago during the industrial revolution were as much precipitators of today's water crisis in the Llano Estacado as were the farmers who settled the area. Indeed, the legacy of cotton is long and rich, fraught with the best and the worst of human character and action. Slavery in the American South, persecution and liberation of women and children in the textile mills, technological discovery, and the "rise of the factory" are just a few of the elements linked to the rise of cotton as a principal commodity crop and fibre (see for example Hareven and Langenbach, 1978; Wright, 1978; Genovese, 1965; Young, 1903; Robinson, 1898). In fact, "King Cotton" and the textile industry were a fundamental part of the first industrial revolution and, along with other elements, led to the integrated world market that was largely in place by 1914.

At the end of the eighteenth century, cotton accounted for only 5 per cent of global textile consumption; by 1890, its share was over 80 per cent. For more than a century, the international division of labour involved the demarcation between raw cotton producers in the colonial and semi-colonial regions and industrial processors in the centre. Major deviations from this simple schematization existed, however. For example, the United States was one of the largest cotton producers and exporters long after it had become a central industrial economy with a major textile and clothing industry – in fact, the

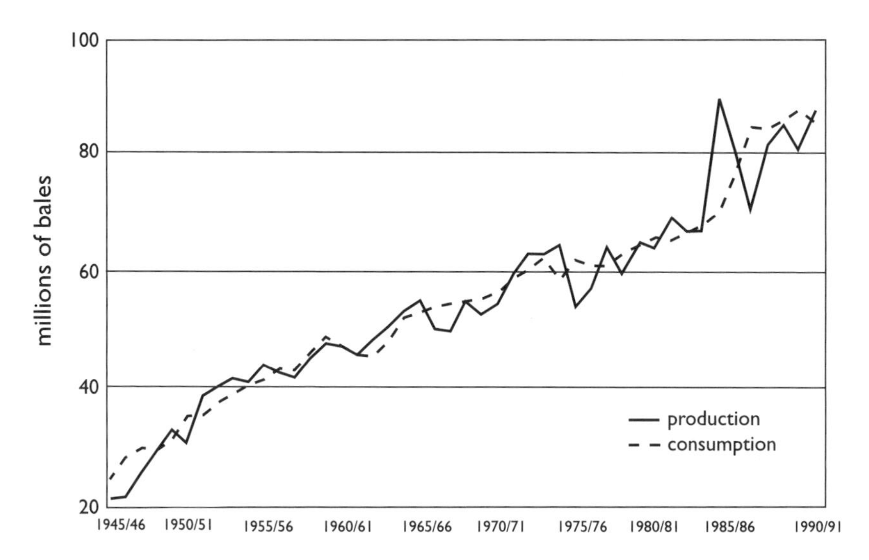

Figure 6.1 **World cotton production and consumption (1945–1991)**
Source: Morris, 1991: 6.

United States was the biggest raw cotton producer until the 1980s, when China overtook it.

The international division of labour with respect to textile manufacturing began to change prior to 1914 as expansion of the textile and clothing industry occurred in several developing countries, notably India, China (later the People's Republic of China or PRC), Brazil, Argentina, Chile, Mexico, Egypt, and the Sudan. Hong Kong, Taiwan, Pakistan, Turkey, Thailand, Indonesia, and Japan became important spinning countries following World War II, with India and the PRC maintaining a large capacity.

World cotton production and consumption have grown steadily since the end of the World War II (figure 6.1). World production rose from 21.4 million bales in the 1940s to over 87 million bales in 1995/96 (Anderson, 1997: 49; Morris, 1991: 6). Over the last 50 years, world cotton planting has ranged from a low of around 22.3 million hectares (55.75 million acres) in the 1945/46 crop year to 34.8 million hectares (87 million acres) in the 1990/91 season (Morris, 1991: 23). For 1994/95, it was estimated at over 32.6 million hectares (81.5 million acres) (Anderson, 1995: 24).

Individual cotton growers are increasingly at the mercy of the international market, in addition to the constants of weather and insects. Technological developments, weak income elasticity of demand compared with manufacturing, government supports in many countries, high ratios of carryover stocks, and other factors have tended to keep the price of cotton low relative to those of manufactured products. Like the prices of many other industrial raw materials (Radetzki, 1990), cotton prices have tended to trend downwards in real terms. By the end of the 1980s, real value prices were only a third of what they were in 1945/46 (Morris, 1991: 6). Wide price fluctuations, another characteristic of most commodities, are expected to increase for cotton as deregulation and liberalization continue apace.

Production varies considerably from year to year. Cotton is a crop that produces high value per hectare, is non-perishable, has low weight to volume for transport, and can survive drought to some degree. Many countries depend upon it for foreign exchange and have tried expanding its production over the years. Farmers in the United States and elsewhere increase cotton acreage when the price goes up. The crop is also susceptible to damage from insects and rusts, and can be affected dramatically by climatic variability. All these factors, in combination with technological advances, the prices of other crops, government policies, and the prices of inputs, create flux in production.

Yields roughly correspond to inputs, which are, in turn, roughly reflected in costs of production. These vary considerably depending on farm size, level of mechanization (including whether cotton is seeded and picked by hand or by machine), irrigation method used, quality of chemicals applied and variety grown. In the United States, where the average cost of production was US$1,137/ha ($460/acre) in 1988, costs can vary from a low of $568/ha ($230/acre) in the southern plains to $1,977/ha ($800/acre) in the far west (International Cotton Advisory Committee, 1988).

World cotton production is not expected to increase much in the medium-term future. Cotton is a relatively high-risk crop and many US growers, in the absence of price supports and subsidies, are unwilling to plant it, especially when other commodity crop prices are attractive. Furthermore, in many countries, cotton competes with food crops for land, constraining its expansion; current technology is likely to result in only modest gains in cotton yields in the near term.

Cotton exports and trade

Cotton has been traded internationally for over two centuries. In 1850, nearly 90 per cent of US cotton was exported, with earnings equal to about two-thirds of all goods imported into the country (Sanford, 1990). In 1988, almost 40 per cent of US cotton was exported, with earnings equal to less than 0.5 per cent of all goods imported into the United States. Even so, the US cotton trade generates a surplus of nearly $2 billion. The grower's profit, once the cotton is brought to market, depends upon the dollar exchange rate, cotton production in other countries, import and export policies of other countries, existing surpluses or stocks worldwide, and the demand for cotton.

Approximately 30 per cent of world cotton production is traded internationally. Over the past three decades, as output has expanded, so has the volume of raw fibre traded. Table 6.1 provides an estimate of the trading volumes for major cotton producing countries. Six of the largest transnational corporations account for between 85 and 90 per cent of internationally traded cotton, usurping the growers' influence in the trading circle (Maizels, 1992). The cotton grower's price to the trader represents only 4–8 per cent of the final product (cotton clothing) price (Maizels, 1992: 164).

Most of the Llano cotton crop (60–70 per cent) is exported for use in making lesser quality yarns typical of those used in the denim market. In effect, the Llano competes with Pakistan, Uzbekistan, the PRC, and other areas that grow cotton of a similar quality. Reliability (of delivery and quality) is one of the big competitive advantages the region has vis-à-vis its competitors. The North American Free Trade Agreement is expected to benefit Texas growers over the long term because cotton can be delivered to Mexican mills quickly and efficiently. In addition, new mills are being planned for state-of-the-art cotton spinning in Texas. Whether these new opportunities will aid the region's cotton growers is uncertain, primarily because the costs of production have edged upward over the last few years. The region went from having relatively low-cost cotton production in the 1970s to high-cost cotton production by 1985 (Bednarz and Ethridge, 1990).

These increases in production costs (including higher irrigation outlays), along with greater competition in the global market and government deregulation, have caused a continued decline in the number of area farms and in farm sales (see figures 1.2, p. 16, and 6.2). Average farm size tended to grow a bit in some counties but, for the most part, stayed relatively constant from the 1970s to mid-1990s (see figures 6.3 and 6.4). The downturn or cost–price squeeze prompted many Llano farmers to go to a combination of tillage

Table 6.1 **Exports of cotton by country/region (1980–1990) (thousand bales)**

	1980/81	1981/82	1982/83	1983/84	1984/85	1985/86	1986/87	1987/88	1988/89	1989/90
USA	5,926	6,567	5,207	6,786	6,215	1,960	6,684	6,600	5,900	6,983
Brazil	137	828	78	354	413	357	304	560	350	400
Egypt	782	899	946	845	667	662	652	436	550	475
USSR (primarily Uzbekistan)	4,023	4,275	3,996	3,283	2,984	3,137	3,344	3,200	3,000	3,200
PRC	6	0	75	760	944	2,822	3,169	2,100	1,750	1,750
India	566	275	553	288	122	286	1,224	275	300	450
Pakistan	1,504	1,069	1,222	377	1,260	3,146	2,893	2,500	2,733	3,300
Mexico	802	759	359	434	574	444	220	364	400	277
Turkey	1,020	956	730	499	686	304	509	250	400	450
Australia	209	254	238	332	483	441	448	576	561	561
West Africa	818	702	928	1,035	1,318	1,358	1,589	1,778	1,738	1,869
World	19,922	20,977	18,563	19,990	20,761	20,507	26,666	24,060	22,823	25,500

Source: Bulletins of the International Cotton Advisory Committee.

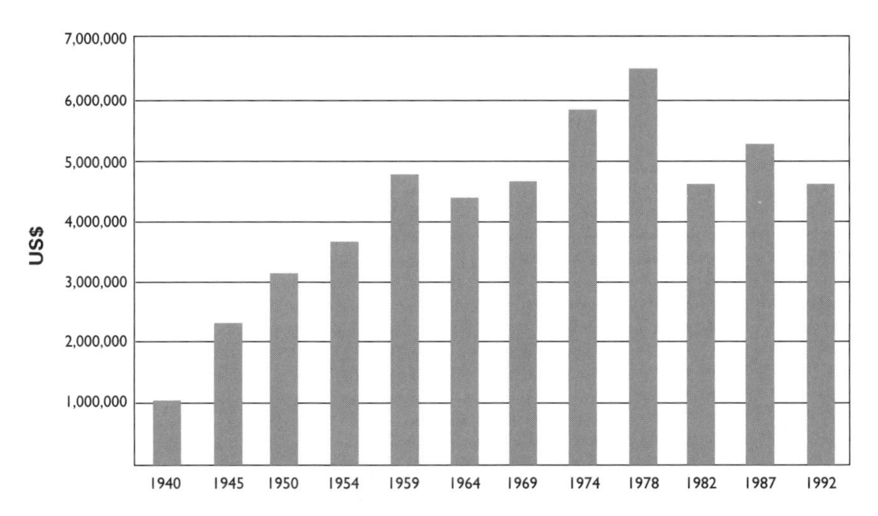

Figure 6.2 **Market value of agricultural products sold: Selected counties of Texas–New Mexico study area (adjusted by CPI to 1995 dollars)**
Source: US Agricultural Census.

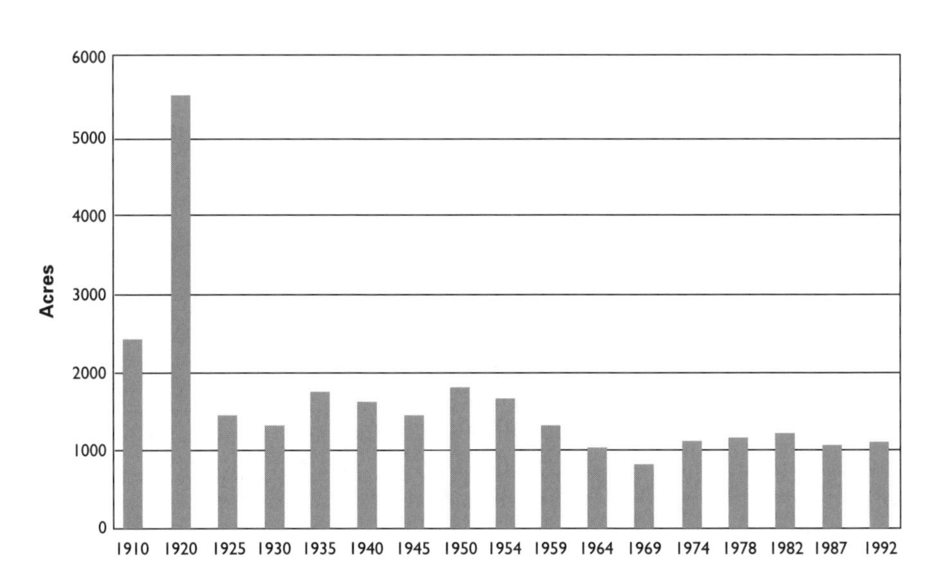

Figure 6.3 **Gaines County, Texas: Average farm size (1910–1992)**
Source: US Agricultural Census.

treatments and chemical applications to avoid skyrocketing diesel fuel costs. Growers are also using more efficient equipment and many types of water conservation techniques (row dams, centre pivot irrigation, furrow dyking, surge and drip), and have adopted other cost-

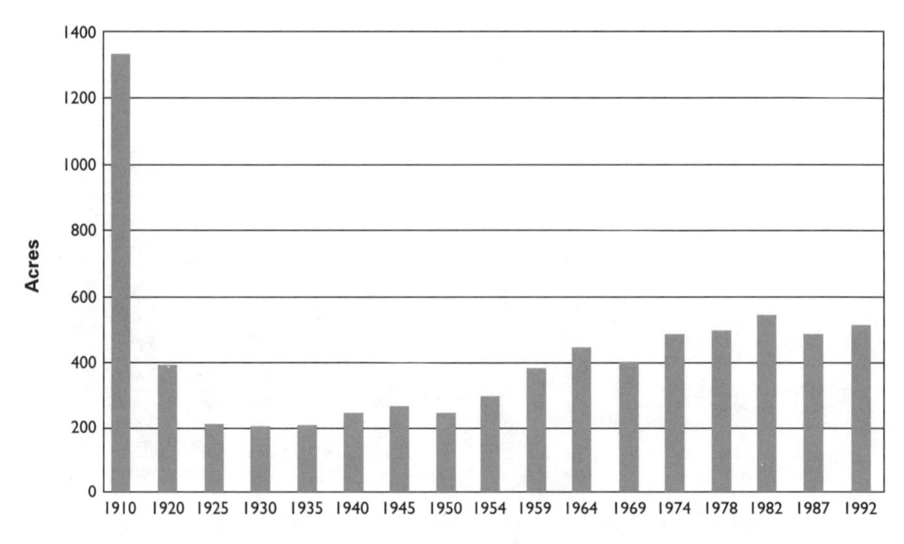

Figure 6.4 **Lubbock County, Texas: Average farm size (1910–1992)**
Source: US Agricultural Census.

efficient measures. Planting new transgenic cottons that are toxic to some pests and tolerant of new pesticides represents another round in creating new margins through technology.

Because nearly all cotton trade is for future delivery, usually within three to twelve months, the futures exchange provides the major market for establishing world cotton prices. The major futures market is the New York Cotton Exchange, established in 1870 to offset price movements. Although speculative trading now dominates the New York Cotton Exchange, its original role has not been lost (Morris, 1991). Many of the farmers on the Llano use the market to hedge their growing and marketing practices; it is touted in the trade magazines as the most important risk management strategy available to farmers now that US farm policy has drastically altered price supports for cotton.

Farmers have always sought to combat uncertainty: only a very few variables are amenable to adjustments in the cotton culture of the Llano. Water supply was thought to be the answer to uncertain weather. Federal price support programmes and scarcity policies were the answer to shifting markets and demand. Now that farmers are facing the limits of the water supply and the shrinking of government support dictated by the 1996 Freedom to Farm Act, they are looking to other means of risk minimization. The future of cotton cultivation on the Llano is dependent on their success. Yet US growers are much better disposed toward dealing with the uncertainties of a global market than are growers from many developing countries. US growers are more sophisticated in terms of marketing, and they have at their fingertips a considerable technological and informational infrastructure. So, for the groundwater aquifer, the implications of deregulation are not necessarily positive. Water levels that held steady or rose over the lean years of the 1980s have, in the 1990s, declined again because of the continuing demand and relatively good prices for cotton. With both dryland and irrigated cotton technology and cultural practices improving, cotton will most likely continue to play a major role in the economy of the Llano Estacado.

Cattle on the post-expansion Llano

Fluctuations in the fortunes and prospects of cattle producers on the Llano were not dampened by the transition to an era that has been defined by the limits of irrigation and roiling international markets. The 1974 drop in cattle prices signalled the start of a major agricultural restructuring. The impacts of the "energy-food-currency shocks" (McMichael, 1993) that hit the US economy were compounded by an increasing health consciousness among US consumers. Consumers shifted their meat preference toward mass-produced and inexpensive poultry, apparently at the expense of beef. US consumption of beef fell nearly 30 per cent from 1976 to 1990, from a per capita consumption rate of 42.9 kilos (94.5 pounds) in 1976 to around 30 kilos (about 66 pounds) through the 1990s (Brooke, 1996; Ufkes, 1995). Declining demand combined with falling profits led to a series of industry restructurings designed to regain producers' competitiveness through economies-of-scale efficiencies and reduction in costs. The most important change in the meatpacking industry was the controversial meatpacking consolidations of the 1970s and 1980s.

Consolidation

A flurry of new players and consolidated old players emerged from these consolidations. IBP (formerly called Iowa Beef Packers) became the most important company heralding the post-1973 beef industry restructuring. Meatpacking competitors became larger, fewer in number, and increasingly vertically integrated, as well (Purcell, 1990a). Old and familiar names in the meatpacking industry such as

Swift, Wilson, Hormel, and Armour became subsumed under large, international food conglomerates. ConAgra purchased Monfort and Swift Independent in 1987. Cargill, the world's largest grain trader, purchased Excel and Caprock industries. The concentration among the "Big Three" (IBP, Cargill, and ConAgra) or "Big Four" (if National Beef is included) reached 69.7 per cent of market share in 1988. In 1989, the Big Four accounted for 80.8 per cent of all cattle purchases made that year (Ward, 1990). In 1975, when restructuring began, the four largest firms had accounted for only 25 per cent of beef production (Ufkes, 1995).

Few can argue with the increased efficiency of the consolidated and high-volume packing plants. Costs of cattle slaughter fall with increasing size of operation and there are indications that the increased efficiencies permitted packers to respond to economic changes in the market more quickly (Purcell, 1990b). In 1995, Excel and IBP, which handled more than 93 per cent of all cattle slaughtered on the Llano, had annual slaughter capacities of 1 million and 1.76 million respectively. Five other packing plants exist on the Llano, but combined they account for less than 7 per cent of the area's slaughter capacity.

Similarly, feedlots or feedyards have increased their capacities to phenomenal levels. In Texas, in 1995, forty feedyards had a capacity in excess of 32,000 animals and another 29 had capacities between 16,000 and 31,999. These feedyards marketed 4.67 million cattle in 1995, or 84.2 per cent of all fed-cattle sold (NASS, 1996). Since the heart of the fed-cattle industry (Deaf Smith, Castro, and Parmer counties) lies on the Llano, the numbers are similar. Of the 1.7 million animal feedlot capacity, all but 109,000 head is in feedlots with capacities of 16,000 or greater.

As packers built high-volume plants for efficiency, their need for a constant, steady supply of quality cattle sparked further vertical integration. A practice already prevalent in poultry farming (both eggs and meat) and dairy processing, vertical integration refers to the consolidation of two or more steps in the commodity production or distribution process by a single company (Goldschmidt, 1978). Vertical integration can be accomplished through single ownership of all steps of a process by a corporation or through contract farming. With contract farming, a processor, marketer, or feed-supply house contracts with individual farmers to produce certain levels of commodities at a set price established in advance of the harvest or, in the case of livestock, before shipping to slaughterhouses (Goldschmidt, 1978).

Through the 1970s, most cattle were bid on by packers at public auctions. Cattle prices were driven up during high demand and driven down with oversupplies. But industrial restructuring meant the demise of public auctions and the rise of direct feedlot purchases. In 1960, only 39 per cent of cattle were sold directly from feedlots to packers (Ufkes, 1995), but by 1989 over 90 per cent were direct purchases (Ward, 1990). Most cattle on the Southern High Plains are sold on the cash market. Increasingly, however, packers have bypassed cash market purchases and have turned to "captive supplies" to meet their need for a constant and steady supply of quality cattle.[1] "Captive supplies" refers to any number of relationships that remove cattle from competitive bidding. Forward contracts, packer-fed cattle, and formula purchases are the most common forms of captive supplies.[2] These arrangements ensure a steady supply of quality-controlled cattle, increasing efficiency for high-volume packing plants, which lose money when kill rates fall much below capacity.

Cattle producers on the Llano remain wary of the market power such extreme industrial concentration gives to packers. Only four companies make over 80 per cent of all cattle purchases nationally; this is probably an even higher percentage regionally. Producers worry that this concentrated power depresses cattle prices, particularly through captive supplies, although no hard evidence exists to support this concern (Ward, 1990: 101). However, industry analysts consider the use of captive supplies threatening to base price development because public information is limited and public agencies are unable to inform the producer sufficiently (Purcell, 1990b).

Meatpacker arguments and statistical evidence to the contrary have not convinced most cattle producers, either. For example, producers cite the fact that IBP made profits of around $256 million in 1995 (Brooke, 1996) while thousands of producers declared bankruptcies or liquidated all of their stock. The question remains largely unanswered: are prices simply determined by relative supply and demand and quality, and is the current crisis just a low point in the cattle cycle?

Additionally, the US Department of Agriculture studies investigating cattle producers' pricing complaints reported that livestock operations and processing plants have created considerable local groundwater pollution. These same studies echoed the sentiments of the small rural communities in the Midwest and Great Plains. Meatpacking plants used to provide middle-class wages, but, with low wages, considerable demand on community infrastructure, and the social needs of immigrants, packing plants are not seen as the economic benefit they once were (see Gouveia, 1994 for a case study of this from Lexington, Nebraska).

Internationalization

The upheavals of the early 1970s also ushered the cattle and feedgrain complex of the Llano into a new era of internationalization and rationalization of markets (McMichael, 1992). Strong state regulation and protectionist policies will be gradually replaced by the expansion of liberal market reform, enforced by the International Monetary Fund and World Bank, supported by a global regulatory framework (General Agreement on Tariffs and Trade or GATT, for example). This "GATT-based regime" has meant that national agricultural deregulation and a further internationalization of food production will become predominant (McMichael, 1992: 344). Despite a number of contradictions, agricultural production and trade are increasingly based on international strategies of out-sourcing, and integration of various livestock sectors across international boundaries.

Mirroring other agricultural sectors, the livestock sector in the United States, which has been predominantly domestic, has recently become international (Gouveia, 1994). The two biggest players are Cargill, the world's largest trader of grains, and ConAgra. Beyond being part of the US beef oligopsony, they have invested in many subsectors of the livestock industry around the world. ConAgra, in addition to expanding production and distribution facilities in Western Europe and Russia, has a hand in nearly every aspect of livestock production from animal antibiotics, feed, fertilizer, to cattle ownership (Gouveia, 1994: 131). Cargill, through Excel, has meat-processing facilities in 55 countries (ibid.). IBP, the world's largest beef producer, has chosen to remain focused on the red-meat sector within the United States.[3]

For US cattle producers, the most important immediate result of the internationalization of the beef industry is that more of their cattle are being slaughtered for export. Beef exports have increased noticeably, measured both as a percentage of production and gross weight, in the last decade. In 1995, exports accounted for nearly 7 per cent of all beef production (more than 900 million kilos or 2 billion pounds), up from only 2 per cent in 1985 (USDA data, cited in Glover and Southard, 1995). Japan, the largest importer of US beef, accounted for more than 50 per cent of US exports in 1994. Other Pacific Rim countries, especially Korea, are targeted by the beef industry for increased exports. Beef imports, now nearly equal to exports on a tonnage basis, have been limited since 1980 due to a meat import law. The US tonnage trade deficit is predicted to decrease further if imports continue to decrease (as they have since 1993) and as exports increase. (On a value basis, the United States has been a net exporter since 1991.)

Cattle producers, feedlot owners, meatpackers, and the USDA all applaud this trend of expanding US beef exports. And continued international expansion may help alleviate depressed cattle prices at low points in the cattle cycle. New markets for US beef, like Japan, which opened to the United States in 1988, have had profound impacts on agro-industrial investments. US cattle feeders altered their fattening procedures and meatpackers built processing plants designed to accommodate Japanese tastes (Ufkes, 1993: 226). Japanese investors quickly purchased feedlots and meatpacking plants in the United States; their overall share of the market remains limited, however. Interestingly, US grain producers (who in Texas are usually also the cattle producers) objected to forcing Japan to liberalize its beef market because Japan was purchasing US grain for their own fed-cattle.[4]

The cattle cycle downturn in the mid 1990s served to remind producers how vulnerable they are to the global market and the climate as well. Cattle prices began falling in 1994 and plummeted in 1996. Cattle producers were hit hard for several reasons. First, many economists perceive a classic case of oversupply as the industry enters a low price phase in the cattle cycle; this was clearly true in the mid 1990s. Producers received relatively high prices in 1992 and 1993 when supply was tight, but in 1994 the number of cattle had increased to its highest level since 1986 (nearly 104 million head) and prices dropped precipitously. Texas producers earned on average $75.60 per hundredweight in 1993; that price fell to $62.20 in 1995 (NASS, 1996). The price fell further to the mid to upper $50 range in 1996. Second, producers would not necessarily sell off their herds in reaction solely to falling prices unless grain prices were also high. Prices for grain, particularly corn and sorghum used to fatten cattle before sale, increased as rapidly as cattle prices fell. For example, corn that sold for under $2.00 per bushel at the end of 1994 increased to $4.99 by May, 1996 before falling again late in 1996. With high grain prices, producers would typically run their cattle on wheat to graze until grain prices fell, but that was not possible owing to the third most important reason, a prolonged drought.

The impacts on the Llano from the mid 1990s drought were widespread and harsh. Farmers in Oklahoma and Texas resorted to ploughing under winter wheat crops because the expense of harvesting the crop would not justify the meagre yields. With grain prices high and cattle prices low, it was costing farmers more to fatten cattle than the extra weight was worth. Massive herd liquidation ensued and the national herd total was estimated to have fallen by at least 2 million in 1996 (Feder, 1996). The Oklahoma Agriculture Commissioner estimated that between 5,000 and 10,000 farming families (out of only 70,000) have been forced to declare bankruptcy as a result (Verhovek, 1996). By the end of 1996, many farmers in Castro County, Texas, had not harvested dryland wheat for five years (Ragland, 1996) and Randall county cattle herds were at 15 to 20 per cent of their normal levels. Because the drought persisted for several years, some farmers, anticipating continued drought conditions, planted more dryland crops (Hacker, 1996). Cattle producers try to ride out the low point in the cattle cycle until cattle supplies are lowered and prices once again move upwards. The overriding concern for most producers is simply one of short-term survival. But on the Llano, a significant subgroup – grain farmers with irrigation – will always stand to benefit in the short term.

Water consumption and groundwater depletion

The combination of the mid-1990s drought and high grain and cotton prices prompted many irrigators to pump tremendous amounts of groundwater to irrigate crops. In 1994 alone, groundwater levels fell on average by 0.6 metres (2 feet) or more in seven HP#1 counties (HPUWCD #1, 1996a). The largest declines were in Lubbock and Hale, two counties dominated by cotton cultivation. Cotton was the single largest irrigated crop on the Llano, accounting for 676,000 irrigated hectares (1.67 million acres) and 2.3 billion cubic metres of water (1.86 million acre-feet) in 1994 (see figure 6.5). Most irrigation occurs within a seven-county area surrounding and including Lubbock. But the farther north and west one goes from Lubbock, the less cotton and the more grain crops are grown. The three main grain crops, sorghum, corn, and wheat, accounted for over 450,000 irrigated hectares (1.12 million acres) and consumed 2 billion cubic metres of

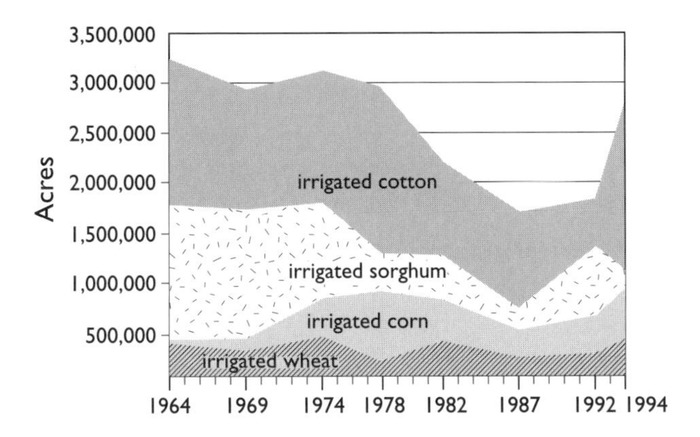

Figure 6.5 **Irrigated acreage: Study area (1964–1994)**
Source: US Agricultural Census.

water (1.63 million acre-feet). All, or nearly all, of the sorghum and most of the corn crop is destined for cattle feed. Some corn in the area, particularly in Hale and Castro counties, is sold to the nearby Frito Lay processing plant. The production from less than about 200,000 hectares of corn (500,000 acres) and nearly 40,500 hectares (about 100,000 acres) of sorghum is not enough to feed the more than 3 million cattle on the Llano.

The Southern High Plains attracted feedlots with its ample feed-grains and cattle, but its success at retaining the feedlot industry and attracting the meatpacking industry has meant that cattle producers from hundreds of miles away send their cattle to the Southern High Plains for grazing, fattening, and ultimately slaughter. Grain production has not kept pace, and much of the corn used to feed cattle is imported from other plains states or the corn belt of the Midwest. Nevertheless, much of the local grain production is irrigated from the Ogallala, and many counties have watched their average depth to groundwater increase by over 6 metres (20 feet). Cotton may be the most important consumer of irrigated water, but irrigated grains and silage for cattle have contributed substantially also, particularly in the northern half of the Llano.

Since the 1940s, some of the greatest groundwater depletion has occurred in the very heart of the fed-cattle industry (Deaf Smith, Parmer, and Castro counties). By 1980, the average depth to ground-water had increased by more than 30 metres (100 feet) in parts of at least seven counties, most notably in Parmer, Castro, Hale, and Floyd (Dugan et al., 1994). Since 1970, Parmer and Castro counties have

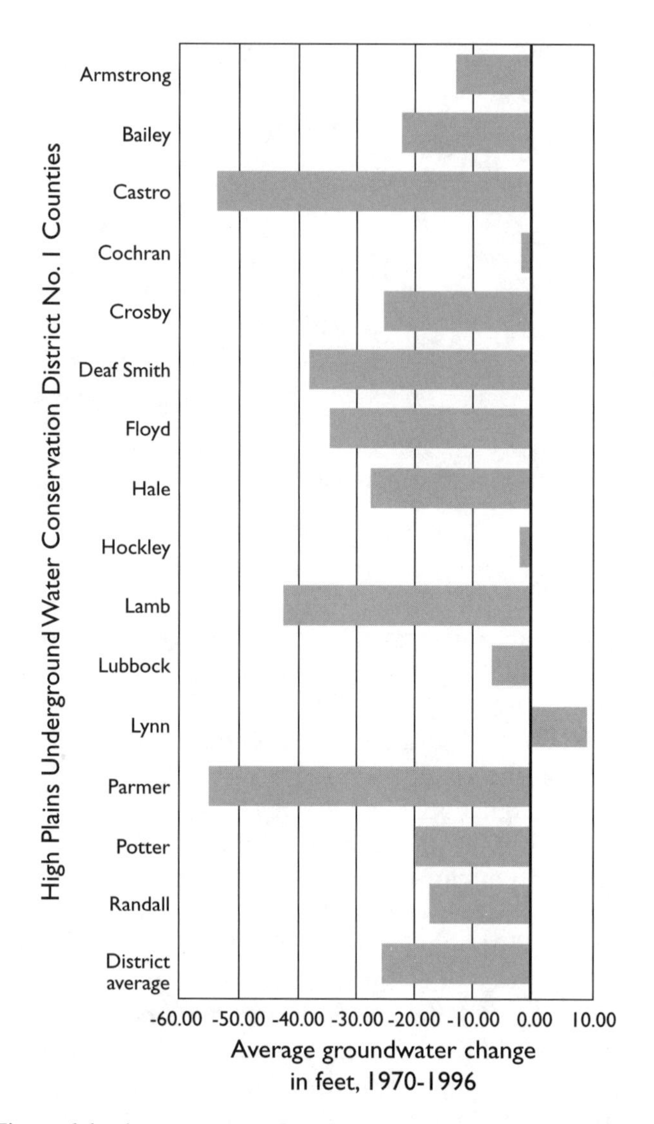

Figure 6.6 **Average groundwater change by county (1970–1996)**
Source: HPUWCD #1.

recorded the greatest increase in depth to groundwater, dropping more than 15 metres (50 feet) (see figure 6.6). Average depth to groundwater has dropped nearly 12 metres (40 feet) in Deaf Smith. These counties are exceptional both because nearly equal amounts of grains and cotton are irrigated,[5] and it was their early irrigation of grains, particularly sorghum, that attracted the fed-cattle industry.

(This trend is also true of Swisher County, but there are no groundwater data available.) In 1958, nearly 98,750 hectares (244,000 acres) and more than 68,000 hectares (168,000 acres) of sorghum were irrigated in Parmer and Castro counties respectively.[6] Given the high levels of irrigation this early, it is no wonder that many people have reached the limits of affordable groundwater, or have continually retooled their equipment to get the same amount of water. Irrigated sorghum acreage has since declined considerably, but Parmer and Castro counties remain leading irrigators of sorghum.

The irrigated land-use figures indicate a diversified irrigated landscape. The land-use adjustment away from irrigated sorghum to irrigated corn has impacts for groundwater consumption. Sorghum, imported to the area from Africa as a drought-resistant crop, is not nearly as water-demanding as corn. In 1994, corn irrigators on the Llano applied 2,529.7 cubic metres of water per hectare (2.05 acre-feet per acre) of High Plains Aquifer groundwater, but sorghum irrigators applied only 1,370 cubic metres per hectare (1.11 acre-feet per acre) (TWDB, 1996b). Very little non-irrigated corn is planted, for a good reason; the Southern High Plains is too dry for such a thirsty crop. If current planting trends continue, correspondingly increased groundwater depletion can be expected.

Fed cattle in feedlots and meatpacking plants consume significant amounts of groundwater as well, but far and away most of the water consumed by the feedgrain–cattle complex is used to grow feedgrains. On the Llano, meatpacking plants consumed 8.3 billion litres (2.2 billion US gallons) of water to slaughter about 3.5 million cattle in 1995.[7] And nearly 3 million fed-cattle in feedlots consumed around 34.4 billion litres of water (9.1 billion US gallons).[8] In absolute terms, this is a considerable amount of water, but converted into irrigation measurements (cubic metres or acre-feet), the meatpacking industry consumed only 8.3 million cubic metres (6,734 acre-feet) and feedlots only 34.6 million cubic metres (28,061 acre-feet) of water. Compared to irrigation, this is nearly trivial: combined, this figure is about the amount used to irrigate grain sorghum in Deaf Smith county in 1994 alone.

Determining the total water consumption of the feedgrain–cattle complex is difficult. Beckett and Oeltjen (1993) estimated that for every kilogram of beef produced in the United States, 3,682 litres (981.9 US gallons) of water were consumed. (They determined that regional disaggregations were not practical because feedgrains that contain water, in addition to being irrigated heavily, move continuously around the country.) This figure is likely to be considerably higher for the Llano because of the dry, hot conditions, causing greater heat stress on cattle and greater reliance on irrigated feedgrains for fattening.

Implications for the regional economy

What does all this mean for the future for agriculture on the Llano, specifically cotton and the feedgrain–cattle complex? There is ample evidence to allow us to conclude that in some places groundwater depletion has led to dryland farming. Emel and Roberts (1995) have noted, from transects in several Llano counties, that there has been a shift from irrigated cropping to rangeland, winter wheat, dryland crops, or idle land, chiefly as a result of shallow soils and a loss of saturated thickness of the aquifer. Anecdotal information supports the idea that many farmers are switching to dryland crops (Forest, 1996) because they can no longer afford to pump what water remains beneath their land. Some older farmers who have experienced the switch to irrigation from dryland agriculture in the 1950s or 1960s are now switching back to dryland crops as their available groundwater is depleted (ibid.).

White (1994) cautions, however, that the entire Great Plains is not the "failed region" some authors would have us believe. Rather, the Great Plains is heterogeneous, and many areas have readily available irrigation groundwater. If "oases" of development based in part on proximity to groundwater develop (or continue), as White predicts, what will happen to the cotton and feedgrain–cattle complex of the Llano Estacado? The answer to that is based on a number of factors, including the severity of groundwater depletion, affordable technologies to extract and use groundwater efficiently, hybrid grain water demands, and the spatial rootedness of the feedlot and meatpacking industries.

Feedgrains grown on the High Plains of Texas consume enormous amounts of water, but much of the grain fed to cattle on the feedlots of the Llano now comes from the less irrigated Midwestern and northern plains states that have not depleted as much of their portion of the High Plains Aquifer groundwater. If less land can be affordably irrigated on the Llano for feedgrains, will the feedlots and packing plants move back to the Midwest, or would such a move be prohibitively expensive? Kromm and White (1992b) argue that more feedgrain crops will be grown in Nebraska and, consequently, feedlot expansion will occur there. Modern feedlots and meatpacking plants, however, represent such large investments that they are effectively immobile. Meatpacking plants, it is also argued, would not move even if feedlots expand in Nebraska because they are close enough (Kromm and White, 1992c: 225).

Amarillo may be a centre for economic growth in the future and there probably will be "oases" of development on the southern plains, but it appears certain that fewer and fewer people will be living as farmers and ranchers. Each cattle cycle pushes more farmers and ranchers out of agriculture. This may produce an economically efficient beef production system, but it may also mean a depopulation of the area as cattle inventories and grains are controlled by fewer people.

One unexplored possibility remains the so-called "designer beef" market – the demand for organically fed free-range cattle, while small, has grown steadily over the past decade. Luxury markets, particularly in Asia and Europe, are seen as likely consumers for the expensive beef. This small niche market will not substitute for the demand that fuelled the giant feedlot–meatpacking enterprises but could provide an outlet for small-scale, less water-intensive cattle operations.

Current water use and water management in the Conservation District

Groundwater levels in the High Plains Aquifer within the High Plains Underground Water Conservation District No. 1 (HP#1) service area stabilized in 1985, for the first time since widespread withdrawals began. Since that time, depth-to-water measurements have revealed rises in some years while other years have shown declines in groundwater levels within the district. In the 1990s, measurements revealed an average decline over all years but for 1992. The drought that began in mid-1992 contributed to the increased rate of decline in recent years. From 1991 to 1996 the average decline over the area was 0.375 metres (1.23 feet) (HPUWCD #1, 1996b: 1). During the extended episode of drought beginning in 1992, the average annual rate of net depletion of the aquifer in the 15-county HP#1 service area was 2.64 times greater than the annual net depletion rate for the prior five years (ibid.). Some areas of the district have experienced water level declines in excess of 30.5 metres (100 feet) from predevelopment to 1990. Texas Agricultural Extension Service specialists estimated in May 1996 that 90 per cent of the dryland cotton crop was not planted due to drought conditions (HPUWCD #1, 1996c). The severity of that drought led to legislative initiatives to limit non-essential uses of fresh water, for example, the use of fresh water in oil and gas recovery.

Short of importing water, groundwater decline in a virtually non-renewable aquifer is an intractable problem. Only a few alternatives are available to water users, and these are being considered in the Llano Estacado because of the mid-1990s drought conditions. As groundwater levels in the aquifer continue to decline, many municipal, industrial, and agricultural water users are looking to other, deeper aquifers for their water needs. These aquifers have less desirable water quality than does the overlying Ogallala formation, but in some places the water quality is marginally acceptable for drinking and irrigation purposes. In 1993, the Edwards Aquifer Authority was legislatively created to deal with issues of rights to use and manage one of these deeper aquifers, with the goal of eventual more widespread extraction. Maximum annual withdrawals were decreed (not to exceed average natural recharge, but adjustable to weather conditions) and regulations on use established, with permits tied to historical uses.

Again, the promise of another endless supply of water has been tempered by the reality of access and usability. While problems of poor water quality and great depths to water associated with these aquifers can be somewhat mitigated by technology, the impact of withdrawals on the environment are less easily dismissed. Downstream impacts of withdrawals were particularly scrutinized in the context of legislation created to protect endangered and threatened species of animals, birds, and plants. The US Fish and Wildlife Service determined minimum streamflows to maintain habitats for certain threatened species of fish and salamanders, limiting the withdrawal rate for upstream users of the Edwards Aquifer System. Once again, there are no easy answers to the question of where the water will come from in times of shortage for the people of the Llano.

Another possibility is precipitation enhancement. Representatives from North American Weather Consultants of Salt Lake City have studied the area to determine the potential of increasing precipitation through seeding clouds with ground-based silver iodide generators within HP#1. In March 1997, the district decided to solicit bids for a spring/summer precipitation enhancement programme, which met with moderate success; preliminary analysis of the data from the 1999 spring and summer season revealed a possible increase of 2–3 inches in the targeted area's long-term precipitation average (HPUWCD #1, 1999).

More conventional water-saving practices have continued to spread throughout the district over the drought years as well. During the past 10 years, thousands of centre pivot sprinkler systems have been put into operation within the 15-county HP #1 service area. Over twice as many centre pivot systems existed in the district in 1995 as did in 1990 (Moseley, 1996: 1–3). The district continues to work with other agencies such as the USDA Agricultural Research Service and the Texas Agricultural Experiment Station to improve irrigation scheduling systems, conservation tillage practices, irrigation efficiencies, among others. Research scientists at various universities in the area are working on plant genetics that will produce more drought- and cold-resistant cotton and other crops.

The most proactive method of fixing a sure water supply for the region has been the creation of water management plans. There have been six attempts at comprehensive state-wide water management plans over the second half of the twentieth century. The first of these was created after extreme flooding ended the severe drought of the mid 1950s. The latest plan for Texas (*Water for Texas: A Consensus-Based Update to the State Water Plan*) was drafted in 1997 and mandated the creation of regional water plans, developed and administered locally. The Ogallala Regional Water Planning Group was instituted to develop a water management plan for the Llano, demarcated in the state plan as Region O: The Panhandle Region. The planning group's agenda involves identifying water demands; detailing water supplies; matching water demands with supplies and identifying problem areas; looking at different organizations dealing with water and the legal considerations concerning laws and regulations; and pulling together information that can be used to conserve water over the short term (0–30 years) and for the long term (30–50 years). The purpose of the plan is to develop and implement water conservation and augmentation programmes that will stabilize the water supply for the region while providing opportunities for growth and development of the towns, cities, and industry in the region, as well as providing for increased economic activity in the agricultural sector (TWDB, 1997; HPUWCD #1, 1998: 3).

About 150 industry and municipal representatives, water districts, and other entities are involved in the Regional Water Planning Group throughout the 46 counties overlying the Ogallala aquifer.

Water banking and inter-basin transfers are also part of the new developments intended to resolve the recurrent water problems in the region. The emphasis, however, is on increased efficiency and improved water-use practices; even with recent data revealing slightly greater recharge potential than previously thought, the water is still a finite resource on the Llano.

How much water is actually left? Models show that there were 514.99 billion cubic metres (417.34 million acre-feet) of recoverable water stored in the High Plains Aquifer System of Texas in 1990 (Peckham and Ashworth, 1993). In 2000, model projections estimate there will be 470.28 billion cubic metres (381.10 million acre-feet) (ibid.). Average irrigation across the region is about 5–7.4 billion cubic metres (4–6 million acre-feet) per year (Urban, 1992). This would suggest that there is at least 50 years' worth of water left in the aquifer, depending upon fuel prices and reasonable well distributions. The population is expected to grow in most of the major urban areas over the next 50 years, so there will be some movement of water away from the agricultural sector toward the municipal and industrial.

Preserving the water: comparisons of institutional forms

Can the differences in groundwater management institutions across the western United States and elsewhere have an impact on the life of the resource (Emel and Roberts, 1995; Moench, 1994; Smith, 1985)? For several years from the early 1970s through the mid 1980s, a considerable literature arose devoted to the problem of the best form of groundwater management. Most of the discussion, at least in the academic literature, centred on the issue of efficiency, particularly as this was defined by welfare economists (see for example Friedman, 1971). Nearly everyone agreed that well-spacing was an important practice – wells pumping water too close together draws down the water table needlessly and ruins the potential for water production over the lifetime of the well. Beyond that, however, a considerable difference of opinion reigned. These opinions can be grouped into those in support of (1) a community-organized management regime, (2) a centralized-state regime, and (3) an unrestricted private-property regime. Public debates on these management alternatives in California, Texas, Nebraska, Arizona, and Kansas – all large groundwater users – coincided with an environmental movement that called for regulatory reform. Municipal demands were growing in many of these states, and the agricultural boom of the early 1980s exacerbated perceived scarcities as well. These debates resonated with the rhetoric of ideological positions and empirical experience and were punctuated by such emotive terms as "big government," "capture," "market efficiency," "planning efficiency," "local knowledge," and "accountability" (Emel and Roberts, 1995).

Controversy over the "best" institutional form for managing groundwater resources reflected deeper concerns over the control and allocation of natural resources (Francis, 1990). During the 1960s and early 1970s, proponents of "regulatory reform" sought to make regulation more effective by preventing agency "capture" by those who were the target of regulation. Critics of existing state and local government argued that most social regulation was "weak, understaffed, and unduly inclined towards cooperative rather than coercive modes of enforcement" (Bardach and Kagan, 1982: 5). Regulatory reform thus implied tougher, more centralized, less discretionary, and more rule-oriented regulation. Many favoured the idea of a single, powerful state-wide super-agency with clearly defined authority, an ample budget for its tasks, and the expertise necessary for rationally planning and managing water resources for quantity and quality (Bradley et al., 1984). These policy analysts stood in the tradition of earlier American Progressives and later "ecologists" who argued for a Weberian model of state-administered rights that would reduce the possibility of capture or cooptation (Young, 1982; Emel and Brooks, 1988).

In the early 1980s, "regulatory reform" took on a different meaning. Proponents sought to moderate "overregulation" and its excesses (including the costs to producers), occasioned by the tougher protective regulation of the 1970s (Bardach and Kagan, 1982). "New resource economists" such as Terry Anderson (1983) argued for a more narrowly defined regime based on private property and the market. Taking a libertarian tack, these theorists maintained that "big government" was unnecessary for desirable outcomes in resource allocation and use. Along with proponents of common-property institutions (or locally shared, locally determined regimes), libertarian economists maintained that knowledgeable private or community property holders would exercise their rights in such a way as to avoid the "tragedy of the commons" (Dryzek, 1987). One of the most vocal proponents of the populist form (or "self-organizing" form), Eleanor Ostrom, wrote that the actual performance of these institutions depends upon the "attributes of the resource, the local community, and the specific rules used" (Ostrom, 1994: 15).

In sum, new directions and redefined old ways marked the decades following the mid-century boom years of exploitation and expansion. Limits to groundwater use emerged, both economically and practically, as depths to water increased and crop and cattle prices faltered. Ever higher energy prices also limited the profligate use of irrigation as pumping and application costs continued to climb. Regulation of groundwater use increased incrementally as markets and costs did the real work of managing the water. New markets opened up and new ways of producing the commodities for them followed. But the dream of a non-limiting source of water faded as importation schemes were reckoned. Improving efficiencies, strategizing, and scaling back have emerged as the principal means of maintaining an existence on the Llano.

Notes

1. This is particularly true of IBP.
2. A forward contract simply means that a contract is struck between the packer and a producer to purchase cattle at a future date, depending in part on the futures market. Packer-fed cattle means that packers either feed their own cattle or pay to have cattle custom-fed in private feedlots. Formula purchasing is a complicated agreement between the packer and the producer whereby a formula is developed for the value of the animal, based on the cash market value of cattle and the grade and yield of the meat. The specifics of the agreement are restricted to the parties involved. Most contracting is done in large feedlots (capacity exceeding 16,000), where most cattle are fed in Texas (USDA, 1996d; Ward, 1990).
3. See Ufkes (1993) for a discussion of opening the Japanese market to US beef exports and shifts in "agro-food capital" as a result.
4. The USDA is seeking other overseas markets for US beef. European Union markets remain closed to US beef imports; differing standards for safety and healthfulness, particularly the acceptable levels of growth promoters in imported beef, have led to a dispute being heard by the World Trade Organization. As of this writing, the EU has reaffirmed its ban on US hormone-treated beef, citing health concerns.
5. Deaf Smith does not fit this trend as neatly because cotton cultivation is limited.
6. Hale county irrigated more sorghum than any other county in 1958, but its wheat and sorghum irrigated acreage is relatively low compared to those of Parmer and Castro counties and fed-cattle numbers are considerably lower. Average depth to groundwater, however, has fallen by over 8.5 metres (nearly 28 feet) since 1970.
7. Water consumption figures come from the packing plants, which submit totals voluntarily to the TWDB. The slaughter figure of 3.5 million is a calculation assuming a 90 per cent slaughter of capacity. Individual packing plants are hesitant to discuss specific figures.
8. The figure of 11,355 litres (3,000 gallons) per animal was based on sample figures submitted by feedlots to the TWDB. Fed-cattle figures come from Southwestern

Public Service Company, and their numbers are somewhat low because they do not release information on counties where reported numbers would disclose individual operations. Feedlots were assumed to operate at 95 per cent capacity in order not to overestimate feedlot consumption. This is, however, probably a low estimate of water consumed directly by cattle in feedlots (excluding water in feedstuffs), but even if the estimate is off by 100 per cent, the point is the same: fed-cattle consume a significant amount of water before slaughter, but it is a very small amount compared to irrigated feedgrains.

7

The future for the Llano: Realms of necessity and freedom

> The modern critique drew its energy and its legitimation from the unshaken belief that a 'solution' can be found, that a 'positive' programme is certainly possible and most certainly imperative. (Bauman, 1995: 21)

The processes that gave rise to the irrigation-dependent cotton and cattle culture on the Llano were numerous – scientific, entrepreneurial, institutional, cultural, and ideological, among others. They operated at many different spatial scales and can be characterized by historical roots of differing depths. As Foucault (1972: 3) wrote:

> Beneath the rapidly changing history of governments, wars, and famines, there emerge other, apparently unmoving histories: the history of sea routes, the history of corn or of gold-mining, the history of drought and of irrigation, the history of crop rotation, the history of the balance achieved by the human species between hunger and abundance.

Any explanation of the path to the Llano's current dependence on non-renewable groundwater must include a number of intersecting histories: those of barter and trade, cotton-growing and fabric-making, futures markets, patenting, slavery, and so forth. These myriad histories woven together can only begin to tell the story of the role of this fibre in producing both wealth and environmental depletion in the Llano.

The current world of the Llano comprises complex commodity production for global markets, and hard-won voluntary water management efforts. The analysis of the forces leading to the region's current state of endangerment, both environmental and societal, also leads to some potential trajectories for the future in the region. Clearly, many forces have conspired to bring the region to where it is. Multiple layers of history, embracing myriad motivational ideologies, technological developments, institutional evolution, economic production and consumption, global trade patterns, colonization, and dreams of wealth – or simple survival – have shaped the current patterns of water and soil exploitation, and what the actual physical environment has become.

And how is this region different from those other places of nonrenewable water resource dependency? Overuse of groundwater is pervasive in parts of India, the Middle East, Thailand, Mexico, North Africa, and other western US states. Saudi Arabia mines groundwater to satisfy 75 per cent of its water demand – largely to become more self-sufficient in food production (Bandyopadhyay, 1989). Libya, building one of the largest engineering projects in the world, brings fossil groundwater from the south to the northern farms and industries. This effort, also funded with oil earnings, is intended to support agricultural and industrial self-sufficiency. In China, water tables are dropping beneath Beijing, Tianjin, and nearly 100 other cities and towns, primarily in the northern and coastal regions. Groundwater mining is ubiquitous in the Deccan Plateau of India. A large number of villages are completely without water and have to walk or truck it in from elsewhere. Israel also has problems with future water supplies, owing to overdrafting existing aquifers. While the implications of these dependencies are context- and site-specific, the problem of societies and natural limits is global; is it also intractable?

Necessity and freedom

During the summer of 1996, the Texas Agricultural Extension Service at College Station issued numerous bulletins for farm and ranch families dealing with drought. A bulletin entitled "Financial strategies for farm families" offered advice on the importance of communicating within the family and with creditors. For example, experts cautioned against cancelling life insurance policies and encouraged stricken families to apprise creditors of the drought in Texas. They also warned families of the added stress for the "second job" earner (usually a working woman), as she tries to bring cash into the family and support her husband, who is "hoping for rain and hoping to be able to make whatever payments are due on the farm enterprise." Older children were encouraged to look for sources of employment (Banks, 1996). This advice makes it easy to understand the continued reliance on groundwater as a source of alleviating hardship. Moving narratives of women hauling water for miles in developing countries inspire instant understanding. People will always live in dry lands with uncertain climates; groundwater development provides security and, often, an improved standard of living. Why, then, leave the water in the ground? Under current conditions in the Llano Estacado, barring any significant above-ground ecological impacts, and given the history of great loss prior to groundwater irrigation, there would appear to be no other solution than to pump the water.

In arid and semi-arid regions, some response to drought or dryness is required to maintain human survival; in wealthy nations like the United States, the economic and technological ability to exploit resources exists. More intangible constraints, political, cultural, and ethical, are often raised. What is possible and what is acceptable by some measure can often conflict. In his treatise on Nature's creation and consumption of the human species, Blackburn (1990: 3) writes:

> Any community is obliged to be prophylactic or recuperative of material destruction, where the damage in question emanates from those rife forces detrimental to its chances of survival. Indeed ... many of the material forms of creative development and not a few of its cultural forms are undertaken in response to challenges or incipient realities of destruction which are posed by other societies or by human involvement with nature.

Blackburn argues that the human need to combat destruction constitutes the "realm of necessity" that exists in opposition to freedom, the "realm" in which societies act, unconstrained. Discerning these degrees of freedom and necessity is a highly controversial undertaking. Just what the possibilities to ensure survival are, and how much they will cost culturally, socially, economically, environmentally, and in terms of liberty, are the issues. And societies do not choose from an endless panoply of possibilities; there are constraints. The specific historical geography and historical demography of a place, along with its other material and cultural characteristics, shape the possibilities that societies have for dealing with the paradox. The behaviour and historical circumstances of other societies, as well as their own environmental pressures, mean that societies must "constantly renegotiate their realm of necessity and the terms upon which they are then capable of autonomous creative activities in their realm of freedom" (Blackburn, 1990: 22).

Two issues persist. First, how are the possibilities for creative action in the realm of freedom identified, acknowledging the historical geographies that led to the present? The second issue, building on the first, has to do with the ethics or morals employed in choosing a path forward. The latter becomes a judgment on the former. What are the moral outcomes of the possibilities that exist for the reproduction of society and nature? What is necessary and what is not? What are humans and their societies capable of? These are the questions that arise within societies, between societies (across space and time), and between societies and the animals, plants, and ecosystems assigned to the "nature" category .

Both problems (identification of possibilities and assignment of ethicality or morality) are ones of interpretation and power. The history of the Llano illustrates this dilemma. The goals of one group of people, a group with the backing of considerable wealth and power, precluded the very existence of the indigenous group. With the Comanches defeated, their population drastically reduced and removed, and the region newly available, the steady work of environmental transformation was accelerated. There can be considerable distance between what *can* be done (and who should or could do it) and what is fair or just. These issues underlie the debate about "sustainable development."

Interpretations of sustainability

We began this book by questioning what is to be sustained on the Llano. Was it the resource, the community, or the ecosystem?, we asked. In the end, what would sustainability mean for the Llano Estacado? Sustaining the resource would entail preserving key economic sectors which would then serve to sustain the region's role in global markets for cotton and cattle. Sustaining the community would imply preserving a way of life and an agricultural production regime. Sustaining the ecosystem is no longer an option. Since the beginning of the twentieth century, the pre-modern ecosystem has been systematically replaced with a fairly precarious one, whose resilience is largely a by-product of huge investments of time, toil, and capital. But if the goal is to sustain the *region*, then the question becomes more complex. A fundamental piece of that puzzle would be to sustain the ability of the people of the region to persevere and transform whatever circumstances evolve into a workable, even if short-term, environment.

The stages of criticality outlined by the ProCEZ researchers could be understood as points of considerable renegotiation of the conditions of necessity and the terms of freedom to prevent what might constitute an irrevocable failure of sustainability. Each stage of criticality, however, might represent less a demarcation than a definable moment on an ineluctable slide through environmental transformation, for good or not, as each did in the case of the Llano.

From the wide-ranging definitions of sustainability, is there a conception of a possible long-term maintenance of production, community, *and* ecology that suggests a future for the agricultural production regime on the Llano Estacado of Texas? If the system, as currently configured, is endangered and therefore unsustainable, what would constructing a long-term human ecology entail? The answers to these questions might suggest a future trajectory for the people of the region and also help to create a definition of sustainability that is both reflexive and practical. Such a definition would then necessarily address that fundamental imperative of sustainability, to maintain something – the groundwater, the history, the way of life – for the longer term, beyond the current human generation.

Gale and Cordray (1994) organized the many definitions of sustainability around three questions: what is to be sustained, why is it to be sustained, and how can this sustainability be appraised? The resultant array ranges from the sustaining of a community, an ecosystem, biodiversity, a global niche, a dominant product, to optimal economic yields. Looking back to the questions of sustainability posited above, that range of definitions might be divided into four perspectives: (1) economics-centred theories of production, (2) human-centred theories of criticality and adaptation, (3) ecocentric theories of preservation, and (4) structurally-centred theories of capitalist production. Each of these perspectives illustrates a possible scenario of what the future will be in a sustainable Llano.

Production-centred theories: Sustaining the resource

The question of measuring what exactly long-term sustainability of resource use would look like has been taken on by resource economists. Using the assumptions of neoclassical and welfare economics, researchers also have attempted to address the intergenerational imperative, that basic premise that we owe something to future farmers and other resource users. The resource economics approach to sustainability can be broken into two central concerns: measuring environmental costs, and evaluating those costs over time. The first issue, recreating the physical environment as an economic model, requires both measuring the immediate production costs of losing important resources and incorporating less tangible costs. Valuation tools have improved considerably in recent years. For example, International Monetary Fund (IMF) economists incorporate resource degradation as a cost of development (Steer and Lutz, 1993). Additionally, work in contingent valuation has improved the understanding of the concept of people's "willingness to pay" for environmental protection and preservation, better informing economists and policy-makers in recent years (Mitchell and Carson, 1989; Pearce et al., 1990).

For the Llano Estacado, the second issue, how to measure costs over time, offers a clearer measure of what the decline in groundwater levels means for costs to the cotton and beef production system. The internalization of these costs is a required precondition for sustainability. Measuring the changing aquifer levels as a tax-deductible depreciation in capital value reflects such an effort (Duncan, 1987). Clearly, the increased costs of pumping caused by the oil crisis in the 1970s resulted in a firmer resolve to conserve groundwater. Assessing some cost to the water beyond the cost of pumping would no doubt result in even more savings and could act as a sort of royalty on the water, much as oil and gas or coal yield royalties to local and state populations. Future efforts might also include some kind of contingent valuation for non-agricultural land uses or non-marketed aspects of the ecosystem lost through intensification.

But how to address that fundamental aspect of sustainability, the need to acknowledge future generations' dependence on the groundwater, which is closer to depletion every day? To preserve some useful life of the groundwater on the Llano, the rate of future preference should be determined, the argument goes, to establish the appropriate incentives for reductions of present extraction rates, slowing the depletion.

Sustainability can thus become a question of measuring the costs, both real and perceived, of protecting resources. Sustainability is the result, according to this theory, of simply achieving the correct measurement of costs. Managing a resource, species, landscape, or ecosystem for sustainability means that the costs to the present must be not only evaluated but also optimized for continued maximization of production. This will not only bring about the preservation of goods for the future but will determine how to balance the benefits of future preservation against the costs of lost productivity in the present. This, the discount rate, or social rate of time preference, should be set so that production and capital reinvestment can be maintained over time (Lines, 1995). For economists, the problem of sustainability then becomes one of setting this discount rate properly, with the assumption that the further in the future use occurs, the lower its current value as a good or benefit (Cline, 1992; Lind, 1982).

This problematic approach raises more questions than it answers. Despite the sophistication of econometric accounting, traditional methods still dominate environmental decision-making and model-building. For example, while acknowledging the importance of alternative indicators, IMF economists insist that "income is still the best measure we have of people's command over many of their commodity needs," and continue to rely on it in environmental evaluation (Steer and Lutz, 1993). On the Llano Estacado, income is not a reliable measure: income from sales of agricultural products has declined in many counties but increased in others, owing to a number of factors. Here the possibility of defining an economics of sustainability is stymied by that criterion, in part because there is no accounting method that can accurately depict the region's heavily capitalized agriculture.

Similarly, the "willingness to pay" principle also draws criticism because sustainability is defined in terms of those benefiting from the "structure of privilege" that determines disposable income (O'Connor, 1994b: 140). Thus valuation is weighted towards political and economic elites. This form of accounting also does little to measure ecosystem or community stability over the long term. Moreover, the problem of the intergenerational imperative remains, despite the economists' tool for acknowledging this, the discount rate principle.

Applying this principle leads to more questions than it seems to answer. First, there is a tendency to estimate the discount rate with marked conservatism in the face of uncertainty. For example, Nordhaus (1994) argues that the best economic policy response to global warming is to do as little as possible. The conservatism of such estimations, it is argued, is inherent in the calculus of a "quantitative logic of capital valorization" (Altvater, 1994), which mandates the neglect of the future. On the Llano, the arguments for undervaluing the future estimation of the groundwater would run to the immediate value to be gained by present exploitation. Additionally, this is a region that has found its future rewritten by technological innovation or government intervention more than a few times. As a result, methods for preserving the water for future agricultural needs are undermined by the belief that the future can only be brighter than the present. The dictates of contemporary growth tend to outweigh any future costs in this sort of cost–benefit analysis (O'Connor, 1994b).

This approach also denies the intrinsic value in the ecosystem because it cannot be satisfactorily measured. On the Llano, the present-day ecology would be valued in terms of production, outputs, market demands; what existed prior to the twentieth century will never be assigned its appropriate value. The notion that a naturally occurring environment can be parcelled out and optimized, it has been argued, is a driving force behind the failure of sustainability (Devall and Sessions, 1985).

For the Llano Estacado, the implications of this production-centred approach to sustainability are clear. Extraction of groundwater and soil must be optimized not only to preserve the resource for the future but, equally importantly, to optimize present production so as not to "rob" future generations of the benefits of contemporary production and reinvestment in technology, equipment, and knowledge. The exploitation of the aquifer in the 1950s and 1960s provided the capital to improve water harvesting and soil preservation techniques and is therefore justified in economic terms despite its heavy toll on the resource base. Further, the reconstruction of the contemporary systems should be as conservative as possible so as not to hamper the efficiency of cotton and beef production in an increasingly competitive global market. This last point is an important one in an economic estimation of sustainability. The consolidation of beef processing and the demise of smaller, less competitive producers in west Texas would be viewed as a step towards sustainability, with higher profits insuring production in the region over time, unless one is interested in sustaining the smaller producers in the name of diversity.

In reality, the historical record in the region belies the assumption of a simple path of capital from production to innovation and back into production. While innovation in irrigation and soil preservation techniques has improved the production regime and its environmental conditions, the bill has been paid largely by taxpayers through the Soil Conservation Service and the Agricultural Extension Service. The capital earmarked, in theory, for innovation has been used predominantly to increase the rate of extraction (Brooks and Emel, 1995).

Human-centred approaches: Sustaining the community

According to the human-centred perspective described by Mortimore (1989) and Bennet (1976), short-term adaptations, even heavily extractive ones like mining groundwater, may create resilience, and thus sustainability, over the long term under conditions of uncertainty. On the Llano Estacado, resilience is an artifact of the adaptive strategies of farmers and ranchers; uncertainty is a way of life. A study of human adaptation on the Llano could detail the use of technology to moderate environmental uncertainty and thus point to a means of ensuring the system over the long term. As Green (1973) points out, the groundwater extraction regime emerged as a stabilizing strategy against rainfall deficiencies in the period following the Dust Bowl. According to an adaptation-based analysis, those that could afford it turned to pumping plant technology to enhance the sustainability of a system that already proved vulnerable in the past, exploiting the stabilizing effect of technological interventions in the Southern High Plains human–environment system.

This approach raises difficult questions, however, and ignores more fundamental barriers to system resilience over the long term. It was not merely the technology of the system or the range of individual responses that was transformed through the implementation of pump irrigation, it was a political and ideological structure that cemented a regime of extraction. Thus a successful short-term solution may become an entrenched way of life. It is also important to remember that those people choosing to irrigate are situated within a social mode of economic regulation – norms, patterns of conduct, procedures, social networks, institutions, and forms of organization – which ensures the reproduction (through its stabilizing and ameliorating of conflicts) of a particular pattern of consumption and production that influences to a great measure what their future choices will be. Though the populist tendencies in this scenario place considerable emphasis on the ingenuity and will of individual farmers, a more structural lens reveals the limitations.

For example, another bulletin from the Texas Agricultural Experiment Station during the 1995/96 drought described the Station's "precision agriculture" programme. The terminology captures the intensive technological approach to plant and animal production that has long been an essential aspect of livelihood on the Llano. Some of the precision agriculture methods include "variable cotton growth regulatory application," "variable rate herbicide application," "decision support software," and "non-invasive determination of yield limiting factors."[1] The tremendous dependency of current farmers and ranchers upon precision agriculture suggests that the rugged, individualistic farmer or rancher is apocryphal. Instead, today's farmers are dependent on a complex of white-coated scientists funded by agricultural conglomerates and huge chemical concerns, implement and seed companies, and well-meaning but myopic bureaucrats and creditors.

Some metaphysicians see technology as autonomous and out of control, "driving civilization to social or literal doom" (Constant, 1989: 426). Neo-Marxists see it as the "pernicious instrument of class oppression" (ibid.), although more frequently, neo-modernists such as Donna Haraway (1997) see science and technology as the only way forward when the issue is control, not modernization. Groundwater depletion on the plains would not have occurred without textile manufacturing machinery, farming machinery, wells and pumping equipment, energy technology development, and the like. But most, if not all, of these technological advances generated positive results as well as negative ones; technological development has ensured greater efficiencies of water use in the region through sprinkler and drip irrigation systems. Yet one might argue that each round of technological innovation, such as the development of Bt cotton – a transgenic variety genetically coded to resist many insects – simply continues the recurrent problem of commodity overproduction. The genius of human adaptability and willingness to embrace technological solutions has led the people of the Llano to the brink of endangerment, all the while pushing what might be inevitable a little further off.

Ecocentric theories: Sustaining the ecosystem

Ecocentric approaches to sustainability arising from ecocentric notions of scarcity and limits were pioneered in the work inspired by *The limits to growth* (Meadows, 1972). Whether advocating the protection of an important niche ecosystem from exploitation (Corry, 1991) or protecting particular land uses against transformation (Bentley, 1984), ecocentric approaches work from the assumption that the "costs" of environmental transformation and damage are far beyond mere declines in direct and indirect use value. The Llano, from this perspective, is essentially unsustainable. The inherent value in the scrublands, the shortgrass ranges, and the roving dune fields was denied in the 1880s. The value inhering in the agricultural lands, urban areas, and cattle range is all too easily measured in terms of yields, real estate, and price per hundredweight. When the discussion centres on sustainability, it is the latter "ecosystem" that is to be maintained. From the ecocentric perspective, however, sustainability is measured through the health, diversity, and vitality of natural systems. The transformation of diverse prairie into monoculture, and the current damage to ecosystem viability from soil loss, soil toxicity, and fertility decline would presumably have been prevented or at least mitigated by a society motivated by an ecocentric view of sustainability. Curiously, the decline of the aquifer, an important indicator for other measures of sustainability, might be less significant from the ecosystemic point of view. The surface ecology evolved independent of the fossil aquifer; the overextraction of the groundwater has an impact in that it makes possible and prolongs the damage done to surface systems through cultivation and grazing. In fact, the eventual decline of groundwater levels past usefulness might be a contribution to environmental sustainability; future agricultural production would be inhibited and limited to dryland systems of grazing and cultivation, generally seen as beneficial to the maintenance of biodiversity).

In its most radical manifestation, this ecocentric approach to the High Plains problem comes in the form of the controversial Buffalo Commons idea (Popper and Popper, 1991). Offering a solution to a wide range of problems across the Great Plains, including depopulation and unemployment, the Poppers proposed an end to agriculture in the region and the reintroduction of indigenous flora and fauna, especially the North American bison. This would create a massive ecological reserve and would entail the continued depopulation of a region that is home to several million people. Other, more market-oriented ecocentric approaches advocate the acquisition of land for private preservation (Matthews, 1992). Such an approach differs from a truly ecocentric one, however, in that the natural environment is regarded here as a consumable ideal, while being protected for its *a priori* and inherent value (Whatmore, 1993; Urry, 1992).

Opposition to ecology-centred approaches to sustainability generally takes two forms. Resource economists view the reservation of resources, in this case, the groundwater, without compensation as a perfect example of the untenable inefficiencies of centralized management. Structural theorists, meanwhile, see the ecologistic responses as naive and impractical. By separating ecology from economy, the driving force of environmental destruction – the creation of value – is ignored, they argue. Preservationist solutions are thus short-term fixes that do not slow the forces of real environmental degradation in the industrialized world, however much they might resolve the localized problem at hand (Redclift, 1991).

Structurally-centred theories: Sustaining the region

Explicitly connecting the condition, quality, and construction of the environment to patterns of capital accumulation, critical theoretical approaches to the question of sustainability offer yet another perspective. A coherent explanation of the nexus between environmental degradation and social reconfigurations is the aim of this approach, while suggesting other analyses of environmental sustainability (O'Connor, 1994b; Redclift, 1991; Fitzsimmons, 1989).

According to this view, the inherent contradictions in the nature–society relationship in capitalism block the path to sustainability. There are two essential contradictions. First, the planet and its resources are ultimately limited and dwindling while capitalist accumulation inevitably grows and expands; biophysical limits and the laws of thermodynamics conflict directly with the investment–profit cycle (Emel and Bridge, 1995; O'Connor, 1988). On the Llano, capital accumulation has run headlong into the limits of the water; traditional impediments like transportation and access to markets have been compensated for with regionalized hubs and distribution networks that overcome the essential remoteness of the plains; compensation for limited water has not been devised yet. Second, the assumption of social and environmental control inherent in capitalism contradicts reality; simply put, capitalism does not direct nature and society the way it does industrial processes. This leads to conflicting aims among capital, nature, and society and the subsequent interjection of compromises, social movements, and other unpredicted outcomes to mediate (O'Connor, 1994b). When confronted with a depleted stock or destroyed resource, capital is used to exploit a new niche or environment, accelerating the decline of other ecological systems. On the Llano, though, there is nowhere else to go, and precious little alternative water to exploit.

This perspective on the role of capital in effecting sustainability provides a new reading of the problem of the Llano Estacado. The emigration of European settlers to the region precipitated the first phase of transformation, replacing the local landscape with a wholly capitalized one in a generation. Still, the dominant institutional structure, the nation state, continued to inhibit accumulation and consolidation during this early period of the 1880s; land distribution policy was designed to promote individual homesteads, for example (Brooks and Emel, 1995). The following years brought an influx of settlers and cattle, only to be succeeded by decline, with drought bringing starvation and out-migration throughout the region. Environmental vagaries and wildly varying rainfall amounts, impeding growth and accumulation, could not be compensated for with increasing capital investment, particularly through the national depression of the early 1890s. Bigger farmers failed on a grander scale; there was no absorption of loss in economies of scale. This pattern was to be repeated throughout the region's history, with an expansion in dry farming followed by the disastrous Dust Bowl era, again followed by the growth of irrigated agriculture. The boom and bust cycle, from the critical theoretical perspective, becomes a cycle of capital accumulation, interrupted by encounters with ecological and social limits, followed, in turn, by a reconstruction of the regime of extraction (O'Connor, 1994b).

When this pattern attracted institutional attempts at mitigation, those attempts were typically isolated moments of activity geared around the resource itself; on the Llano, groundwater conservation districts and water users' associations were ad hoc responses with limited economic and political agendas. The sudden flurry of institutional and political activity on the part of otherwise disinterested irrigators when Mobil Oil sought use of the groundwater to "waterflood" or plunge declining fields in 1984 is another example of a provoked response (Opie, 1993).

Structurally-centred theories predict that the pattern of overexploitation, reformation, accumulation, and decline described by this perspective is ultimately unsustainable and will continue to collide with ecological and social limits. Additionally, government institutions are deeply involved in this cycle:

> The "water buffalo" agencies of federal, state, and local government in the US west respond to drought with ever more extravagant schemes for capturing distant water, rather than considering more appropriate distributions of people and economic activity. (Dryzek, 1995: 181)

Existing institutions have also opted for the maintenance of private property, and through that, private risk-bearing. The institutions of the federal government that offered interventions in the downward spiral of the farmers' stormy relationship with the Llano's environment did in fact go some distance in lessening the overall riskiness of agriculture by keeping farmers secure and on the land. Some of these institutional interventions included crop subsidies, rationalization of planting (to avoid overproduction), crop insurance (against inclement weather), and agricultural extension and farm credit bureaus (to make the new scientific and technological developments understandable and accessible to all farmers). An alternative social arrangement might have undertaken even more sharing of risk, as is done by collective institutions in other parts of the world. If private property had not been so dear to the European settlers, a collective sharing of risk (and profit) might have reduced the need to slaughter 500 million animals and birds on the plains during the latter part of the nineteenth century. Comanche raiders stealing settlers' cattle and horses might then not have created the institutional circumstances, military and federal intervention, for their eradication and removal.

A collaborative view of sustainability

Sustainability becomes more elusive as a definable notion, and more pertinent to the understanding of the history of the Llano. Clearly, human adaptation and ingenuity are part of the answer, but only part. The complexity of federal agriculture policy provides the cost accounting that should point to the correct mix of use and non-use likely to lead to the optimization of livelihoods on the Llano. Obviously, de-population of the Llano and restocking the region with buffalo is not a practical solution – nothing is sustained in that scenario, although the resulting ecological overlay would most likely be self-sustaining into the future. Finally, the analysis of the complicated and uncomfortable relationship of the people of the Llano with the environment, physical and social, reveals a useful way of understanding the past and predicting the future. Exploitation of the groundwater will continue as long as it is practicable. Institutional intervention will be increasingly less aggressive when the water becomes inaccessible. A new relationship will probably be forged, and a future that emerges will probably be familiar, but less so. A way of life will ultimately not be sustained over the long term.

While accepting the intractability of resource depletion, initiative on the part of farmers would appear to contradict the control that resource depletion wields over people's actions. People manage to "get by," and this cannot be ignored by any analysis of the problems of sustaining an endangered or threatened environment. The gamble and risk elements of the production cycle persist now as they did in

1910. Agriculture has remained a "just viable" way of life on the Llano Estacado. The mitigations offered by institutions when that viability erodes a bit further have not been imagined yet; the belief exists on the Llano that when capital accumulation next collides with intractable natural limits, a way will be found yet again to continue on.

Sustaining the region

Where does this leave the people of the Llano today, confronting and coping with a largely unsustainable situation? Judging from the various perspectives on sustainability, we could conclude that the region has become threatened, or endangered, not through failure of adaptation, which from a human perspective has been a very successful strategy, but through the use of natural resources, specifically groundwater mining and consumption of the pre-modern grasslands environment; conflicts with international commodities markets and their relation to other industrial sectors; incomplete solutions of science and technology; political institutions that, along with economic systems, pursue growth; and the lack of political and social will to preserve or maintain biodiversity and "sustainable" forms of livelihood. National and regional commitment to full employment, the family farm, development, labour, industrial and military growth, elimination of poverty, and poverty itself – all of these have also increased pressure on the environment and led to environmental depletion. Many forces lead to non-renewable resource depletion – groundwater mining in many different places has disparate causes – not the least of which is preservation of life, broadly defined. The central question remains: how can moves toward sustainability be encouraged? Moreover, can community sustainability coexist with ecological sustainability and economic sustainability?

Imagining other pasts and futures

The course of European settlement in the region has been driven with the certainty of divine ordination. The particular trajectory of industrialized agriculture could not have been imagined at the turn of the century, but that agriculture would dominate the region was certain. The sense of inevitability that hindsight provides is misleading, however. Even a small change would have sent the course of events in a different direction. If drought had not descended during the

1930s, would the significant role created for federal institutions on the plains ever have evolved? What would the region look like without the price support and commodity subsidy programmes whose roots are buried in the mountain of relief and credit programmes of the Dirty Thirties? What sort of community structure would have developed had dry farming not been unseated by the extreme drought conditions? Or, to go back even further, what if the Comanches had not been driven from their lands?

The descendants of the "hardy pioneers" have become some of the most capitalized and technology-dependent farmers in the world, espousing the "merits of simple living" as practised by their ancestors on the plains "only to be enmeshed by its opposite" (Shi, 1985). Goldschmidt's examination of "what corporate agriculture means for the character of life in American rural communities" (1978: vii) offers much that can be translated for the Llano. In some instances, for example, irrigated agriculture can create a cooperative system. In Goldschmidt's study area of central California farming communities, open-ditch irrigation works were built and maintained communally; on the Llano, each farmer is the owner of his or her own water, pumping plant, and irrigation system. The impacts of the development of large-scale corporate farming on the supporting communities, however, are strikingly similar:

> So the course of Wasco's [California] star was set by the nature of her physical and social environment. Long before the community existed, the agricultural enterprises were established against which her farmers had to compete, and the pattern was set. The very plan of establishing a colony on irrigated lands inevitably called for the production of cash crops at a high cost with abundant cheap labor.... [I]nevitable in an economic sense,... the cash outlay for expensive equipment necessary to pump water meant producing high-value cash crops. And in order to realize the necessary return to cover these costs the new farmers had to compete with established enterprises. Thus they were immediately caught in the established pattern of farming. (Goldschmidt, 1978: 27)

Furthermore, the monocrop agriculture of Goldschmidt's Californian communities clearly matches the cotton/beef-finishing dominance of the Llano. The physical and cultural impacts are similar:

> This tendency to specialize in one or two cash crops has very clear effects upon the social and physical landscape. Basically, it expresses the competition between the old traditional rural values and the urban value system. One of the first evidences of this meets the eye immediately – the virtual

disappearance of the barnyard.... [A] garden is considered a luxury, not because it is work to plant one, but because it is considered cheaper to buy the products at the market and turn the land onto cash crops. ["My husband would plow up that rose garden if I'd let him. He'd plow right up to the bedroom window."] (Goldschmidt, 1978: 28)

An echo can be heard here of the fence-row-to-fence-row ideal expressed frequently on the Llano.

A more subtle effect on the landscape comes out of identifying more closely with business and entrepreneurial attitudes; involvement with investment and return takes precedence over the connection with the land and the seasonal changes we associate with the agrarian ideal. Farming is depersonalized further by the decreasing use of hired labour – large-scale farm machinery has long been the foundation of farming on the Llano. This "industrialization of agriculture ... has not only altered the psychological and social attitudes of the people on the land, it has also affected the people of the town who serve the farm population" (Goldschmidt, 1978: 41).

Typically, farm towns serve their hinterlands entirely as self-sufficient economic units; marketing, processing, ginning, plus all goods and services could be provided in town. Goldschmidt, however, found this "independence" to be only "apparent." Instead, closer examination uncovered the dominance of chain banking, petroleum and fuel oil outlets, "syndicated" supermarkets and drugstores; few truly locally based and locally interested businesses existed (ibid.). On the Llano, this arrangement has further ramifications. The lack of economic diversity in the region heightens its vulnerability to the effects of prolonged drought, or shifts in global markets, or other forces equally beyond the control of individuals. The shallowness of the economy means that excess labour can not be absorbed in other sectors of the economy and, more ominously, that very little new capital finds its way into the region.

These trends toward concentration in agriculture "have slowly and inexorably invaded the local autonomy and independence of the small community throughout America, but most particularly where farming itself is caught in the vortex of industrialization" (Goldschmidt, 1978: 48).

Awareness of the loss of the rural nature and the smaller scale of life in the communities of the plains has also had an impact on the landscape. The desire to recapture something of a distinctly imagined past is manifested in a growing interest in preserving the townscapes of the early part of this century. Almost every town on the Llano has a historical preservation group or, more likely, person, who has undertaken the protection of the main street or a formerly prestigious building, generally a courthouse. Mayors of the larger cities have created commissions to begin preserving remnants of the past in these communities as well. Talk of "heritage tourism" has surfaced, with the idea of creating attractions consisting of homesteads and abandoned towns for people interested in their ancestors and the history of the region. (Some have even described projects aimed at the European descendants of European emigrants to the plains.) Several old hotels have also been restored and made into destination resorts, most notably the old Post Hotel in Post, Texas.

Some of the impetus for this look backward undoubtedly comes from the current status of many of the towns of the Llano. Despite relatively stable population levels throughout the region, the larger towns are growing and the smaller towns are shrinking and sometimes disappearing. With increasing size, the larger cities like Lubbock are experiencing more big-city problems beyond the anomie and class preoccupation predicted by Goldschmidt. Increasing hunger and homelessness has come along with shifting populations and drought. Major new facilities have been built in Lubbock to process food for the network of food banks that operate across the Llano (*Cross Section*, personal communication). Unskilled labour that at one time was the province of seasonal farm work now is a city problem, with the medical centres and service sectors unable to absorb it all.[2]

With this in mind, we turn finally to the issue of what lies in the future for the Llano. Farming clearly will not be a way of life for increasing numbers of people. The urban centres will require an increasing proportion of the area's very constrained water supply. And the likelihood of shocks to the intricately balanced system and surprises in the future in terms of recurrent drought and shifting global demands and markets make prediction nearly impossible. But one thing is clear: after a century of growth and expansion in irrigated agriculture, the tide has slowly and inexorably turned.

Trajectories

What is in store for the Llano? The region's ecology is dramatically different from pre-settlement times, but is also in many ways similar. The principal difference is that grasslands are now cropland or rangeland and these replacement ecologies have water requirements far exceeding those of their predecessors. Prior to settlement, the region was in what can be described as a dynamic equilibrium: vegetation dying back in times of drought, overblooming in times of excess rainfall. Animal species were also highly adapted. People living in the region followed the animals to water, and maintained their own routes and settlements based on access to available water. They were nomadic – they travelled to the water when the water was not available.

Now, and for the past century, much more is dependent on water – crops must be irrigated and feedlots watered. The cities and towns that were built have significant demands for clean water. Large cities have grown: of the 29 counties comprising the Southern High Plains in Texas, only ten have populations over 20,000; of those, only three have populations near or above 100,000. The likelihood of a depopulation of the region to restore the pre-settlement ecology is nearly nil. The question of what will happen with increasing forced or voluntary retirement of cropland is open. Will the lands be rehabilitated in some fashion, or will wind-driven erosion and colonization by weed and pest species claim the land?

The beef industry is a conceivable path to sustaining some of the region's people and communities. Demand for beef has risen from recent lows, but the export market is the most likely avenue for expansion, and other nations are developing their beef exports as well. Possible outlets might include the small niche markets for organically raised and so-called designer beef. But, overall, these seem hardly likely to sustain the Llano economically. As an adjunct to a larger agricultural industry, cattle would fill a useful role; whether that usefulness would offset the water costs is the question.

Cotton will also continue to be part of the region's economy, but most likely to a lesser extent than in the recent past. Downward price pressure may be exerted by developing cotton economies in Asia and the former Soviet republics; coupled with increasing costs for water and other inputs, as well as the removal of federal subsidies for cotton, that may spell the end for some cotton farmers on the plains.

One can only speculate about the effects of a long drought. The outcome of the mid 1990s drought that racked the Llano was a short-term test of the viability of many cotton farmers. Water levels have dropped slightly more each year of this current drought; the increasing lift has been offset by relatively low and stable energy costs. The impacts of a rise in energy prices and continued drought could result in a very different scenario in the near term.

Water is still and always will be the "Lifeblood of the Panhandle Plains." The water is also largely non-renewable; even the optimistic reports of slightly more recharge than has been long estimated cannot change the fact that groundwater mining built the cotton industry and supports the communities and economy of the plains. Choices will be necessary to sustain what is there. Domestic and municipal uses will most likely increasingly edge out agricultural needs – the regional medical centres will eventually take precedence over a cotton farm at some point in the future.

The choice to find alternative water sources has always been pursued. Water importation schemes were permanently vetoed due to their prohibitive costs and their social unfeasibility. But new plans are always being drawn up – plans to tap the underlying aquifers are being discussed and legislated now. Playa storage potential has been modelled for decades, while research into anti-evaporative films has been ongoing. Attempts at precipitation enhancement are a part of the long-term management plan being drawn up for the plains; this is clearly a supportive measure, given the costs and the uncertainty of success. The 1998 approval by the Texas Legislature of a comprehensive water management plan, the 1997 Texas Water Plan, discussed in chapter 6, points in this direction. An interim plan, it will eventually be replaced by one that offers a 50-year scenario of water use and management in the region. Population is projected to rise by 45 per cent over that time period, and water use to drop by over 60 per cent. The largest component of that decline is projected for water used for irrigation (nearly 70 per cent), offset by a perhaps optimistic 68 per cent increase in industrial water use (HPUWCD #1, 1998).

Summing up, looking forward

To close this analysis of a region and its people, it is illuminating to pick up the story of the Comanches. The Comanches ceased to be a presence on the Llano around the beginning of the twentieth century. Nonetheless, a small number of Comanches continue to live in the region, although their world was effectively destroyed when their livelihoods were displaced and their most important resources, land and buffalo, depleted. They have maintained customs and traditions that link them to the past, but also function as their modern identity. The transformed physical environment persists and, with enough applied water, is extremely productive. Nonetheless, the Comanches' world reached criticality, and they disappeared as a force in the region.

The world of the modern irrigation farmers and cattle producers that currently inhabit Comanchería is now endangered. Resources, not buffalo this time, but water and capital, are stretched and are being depleted. Fewer opportunities exist for people to carry on in the livelihoods of their parents and grandparents. But because the modern Llano is intricately linked to the larger context of national and international markets, this world on the Llano will not disappear. The institutions that have bolstered the region in the past will have a role, however diminished, into at least the near-term future. And the culture and traditions of the descendants of the settlers will also persist; their hardiness and embrace of innovation ensures continued investment of work and hope, despite the inevitably decreasing return.

The complexity of moving a society in the direction of sustaining resources, communities, and ecology is the challenge that faces the people of the Llano. The transformation of the physical environment scarcely more than a century ago has been breathtakingly complete. Settlers newly arriving on the Llano in the 1880s dreamed of an agricultural paradise; Comanches watching them arrive most likely had no idea that their world would be altered beyond recognition in less than a generation by those settlers and the momentum of the nation behind them. Sustaining that vision up to today has come at a human and ecological cost too vast for memory to retain and for the imagination to consider; the realm of perceived necessity guided those actions. The Llano Estacado was transformed by perseverance and institutional intervention and technology; the Llano will be sustained by understanding the consequences of acting wisely in the realm of freedom.

Notes

1. World Wide Web at http://ageninfo.tamu.edu/~searcy/task.html. Date: 8 July 1997.
2. Many small towns have invited niche and undesirable industries to the Llano in the name of economic development. These include daylily cultivation, windmill generators, and sludge dumps. Prisons are emerging as one of the few growth industries available in the region: "[S]ome farmers subsidize their farms by working as prison guards" (Bauer, 2000). Few believe that the thriving small-town economies of the mid twentieth century can be re-created with these efforts, however.

References

American News Co. 1874. *New homes, or where to settle*. New York: American News Co.

Anderson, Carl. 1995. New supply signals. *Cotton Grower*, Fall: 25.

Anderson, Carl. 1997. Questionable acres and exports drive market. *Cotton Grower*, January: 49.

Anderson, Terry L. 1983. *Water rights: Scarce resource allocation, bureaucracy, and the environment*. San Francisco and Cambridge, Mass.: Pacific Institute for Public Policy Research.

Anonymous. 1997. Making cotton: News from around the Belt. *Cotton Grower*, January: 8.

Altvater, Elmar. 1994. Ecological and economic modalities of time and space. In O'Connor 1994a: 76–90.

Ball, Charles E. 1992. *The finishing touch: A history of the Texas Cattle Feeders Association and cattle feeding in the southwest*. Amarillo: Texas Cattle Feeders Association.

Bamforth, Douglas. 1988. *Ecological and human organization on the Great Plains*. New York: Plenum Press.

Bandyopadhyay, Jayanto. 1989. Risking Confusion of drought and man-induced water scarcity. *Ambio* 18 (5): 89–92.

Banks, Penny. 1996. Financial strategies for farm families. http://agnews.tamu.edu/stories/DRGHT/finstat.htm.

Bardach, Eugene, and Robert Kagan. 1982. *Social regulation: Strategies for reform*. San Francisco and New Brunswick: Institute for Contemporary Studies.

Bauer, Esther M. 2000. An Earth without a future. *Washington Post*, 28 March 2000, p. A3.

Bauman, Zygmunt. 1995. *Life in fragments: Essays in postmodern morality* Oxford and Cambridge, Mass.: Blackwell.

Beckett, J.L., and J.W. Oeltjen. 1993. Estimation of the water requirement for beef production in the United States. *Journal of Animal Science* 71: 818–826.

Bednarz, M.S., and Ethridge, D. 1990. Sources of rising unit costs of producing cotton in the Texas High Plains. In *Proceedings of the Beltwide Cotton Production Research Conferences, January 9–14, 1990, Las Vegas, Nevada* (Memphis: National Cotton Council of America): 390–393.

Belt, Joe. 1983. *Lubbock life*. Lubbock: SEI Publications.

Bennet, J.W. 1976. Anticipation, adaptation, and the concept of culture in anthropology. *Science* 192: 847–853.

Bentley, F. 1984. Why protect agricultural land? *Journal of Soil and Water Conservation* 39: 295.

Blackburn, Richard James. 1990. *The vampire of reason: An essay in the philosophy of history*. London and New York: Verso.

Blaikie, Piers, and Harold Brookfield. 1987. *Land degradation and society*. London and New York: Methuen.

Blodgett, J. 1988. *Land of bright promise: Advertising the Texas Panhandle and South Plains, 1870–1917*. Austin: University of Texas Press.

Bowden, Charles. 1977. *Killing the hidden waters: The slow destruction of water resources in the American Southwest*. Austin: University of Texas Press.

Bowden, M. 1975. The Great American desert in the American mind: the historiography of a geographical notion. In Lowenthal, David, and Martyn Bowden, eds, *Geographies of the mind*. New York: Oxford University Press.

Bowers, J. 1995. Sustainability, agriculture, and agricultural policy. *Environment and Planning A: International Journal of Unban and Regional Research* 27: 1231–1243.

Bradley, Michael D., Jonathan G. Taylor, Peter Feldman, and Jacqueline Rich. 1982. Water resources policy for Arizona. Unpublished paper presented at the Conference on Water Resource Policy for Arizona: Who Controls It?, Tucson, Arizona, 10–11 May.

Brengle, K.G. 1982. *Principles and practices of dryland farming*. Boulder: Associated University Press.

Brooke, James. 1996. Beef prices for ranchers are at a 10 year low. *New York Times*, 12 May 1996.

Brooks, E. 1996. The Hardy Sons of Pioneers v. The Boys in the Piney Woods: An empirical analysis of ideology and the decision to manage groundwater. Dissertation, Clark University, Worcester MA.

Brooks, E., and J. Emel. 1995. The Llano Estacado of the American Southern High Plains. In Kasperson, Kasperson, and Turner 1995: 255–303.

Browne, W.A. 1937. Agriculture in the Llano Estacado. *Economic Geography* 13: 156–174.

Brundtland, Gro Harlem. 1987. *Our Common Future: Report of the World Commission on Environment and Development*. New York: United Nations.

Cacek, T. 1984. Organic farming: the other conservation farming system. *Journal of Soil and Water Conservation* 39: 357–60.

Catlin, George. 1841. *Letters and notes on the manners, customs, and conditions of the North American Indians*. Volume I. New York: Wiley and Putnam.

Chen, D.T., and C. Anderson. 1990. Cotton market responses to 1985 Food Security Act, dollar devaluation and US weather disturbances. In *Proceedings of the Beltwide Cotton Production Research Conferences, January 9–14, 1990, Las Vegas, Nevada* (Memphis: National Cotton Council of America): 464–469.

Cisneros, George D. 1980. Texas underground water law: The need for protection and conservation of a limited resource. *Texas Tech Law Review* 11 (3): 637–653.

Clairmonte, Frederick, and John Cavanagh. 1983. *The world in their web: Dynamics of textile multinationals*. London: Zed Press.

Cleland, C. 1966. Do we need a sociology of arid regions? In *Social research in North American moisture-deficient regions; A Symposium held during the forty-second annual meeting of the Southwestern and Rocky Mountain Division of the American Association for the Advancement of Science, May 4, 1966, Las Cruces NM*. Arranged by John W. Bennett for the Committee on Desert and Arid Zones Research. Las Cruces, New Mexico: AAAS.

Cleveland, O.A. 1994. Tradewinds: World cotton production recovering. *Cotton Grower*, August: 25.

Cleveland, O.A. 1995. Tradewinds: A sharp ARP increase? *Cotton Grower* August: 25.

Cleveland, O.A. 1997. Tradewinds: Look outside the belt. *Cotton Grower*, January: 8.

Cline, W.R. 1992. *The economics of global warming*. Washington DC: Institute for International Economics.

Cochrane, Willard. 1979. *The development of American agriculture: A historical analysis*. Minneapolis: University of Minnesota Press.

Connor, John M. 1989. Competitive issues in the beef sector: Can beef compete in the 1990s? In Johnson, D. Gale, John M. Connor, Timothy Josling, Andrew Schmitz, and G. Edward Schuh, *Concentration issues in the US beef subsector*. Storrs, Conn.: NE-165, Department of Agricultural Economics and Rural Sociology, University of Connecticut.

Constant, E.W. 1989. Cause or consequence: Science, technology, and regulatory change in the oil business in Texas, 1930–1975.

Corry, S. 1991. The rainforest harvest: Who reaps the benefit. *Ecologist* 23 (4): 148–153.

Cronin, J.G. 1969. Groundwater in the Ogallala Formation in the Southern High Plains of Texas and New Mexico. *Hydrologic Investigations Atlas HA-330*, Washington DC: US Geological Survey.

Cronon, William. 1983. *Changes in the land: Indians, colonists, and the ecology of New England*. New York: Hill and Wang.

Cronon, William. 1992. *Nature's Metropolis: Chicago and the Great West*. New York: WW Norton.

Cronon, William, George Miles, and Jay Gitlin. 1992. *Under an open sky: Rethinking America's Western past*. New York: WW Norton.

Dale, Edward Everett. 1930. *The range cattle industry*. Norman: University of Oklahoma Press.

Dallas, R. 1990. The agricultural collapse of the arid midwest. *Geographical Magazine* 62 (10): 16–20.

Daniel, Tamara. 1995a. Texas High Plains enjoys distinction as a leading cotton producing region. *The Cross Section* 41 (11): 3–4.

Daniel, Tamara. 1995b. High Plains cotton production draws several processors to area. *The Cross Section* 41 (12): 3.
Darrow, R.A. 1958. Origin and development of the vegetational communities of the southwest. In Shields, Lora M., and Linton J. Gardner, eds, *Bioecology of the arid and semiarid lands of the Southwest.* Symposium on the bioecology of the arid and semiarid lands of the Southwest, held at New Mexico Highlands University during the 34th annual meeting of the Southwestern and Rocky Mountain Division of the American Association for the Advancement of Science and the New Mexico Academy of Science, 29 April 1958. Las Vegas, New Mexico: AAAS (Committee on Desert and Arid Zones) and New Mexico Highlands University, 1961.
Davis, Ernest E. 1994. *Texas livestock prices and statistics.* College Station: Texas Agricultural Extension Service, Texas A&M University.
Dethloff, Henry C., and Irvin M. May Jr. 1982. *Southwestern agriculture pre-Columbian to modern.* College Station: Texas A&M University Press.
Deutsch, S. 1992. Landscape of enclaves: Race relations in the West, 1865–1990. In Cronon et al., 1992: 110–131.
Devall, B., and G. Sessions. 1985. *Deep ecology: Living as if nature mattered.* Salt Lake City: Peregrine Smith.
Dillard-Powell Land Co. 1908/1909. *The best cheap lands in the southwest for diversified farming.* Lubbock: Dillard-Powell Land Co.
Dodge, Bertha S. 1984. *Cotton: The plant that would be king.* Austin: University of Texas.
Doughty, R., and B., Parmenter. 1989. *Endangered species: Disappearing animals and plants in the Lone Star State.* Austin: Texas Monthly Press.
Dregne, Harold. 1985. Aridity and land degradation. *Environment* 27 (8): 16, 18–20, 28–33.
Dryzek, John S. 1994. Ecology and discursive democracy: beyond liberal capitalism and the administrative state. In O'Connor, 1994a: 192–197.
Dryzek, John S. 1987. *Rational ecology: Environment and political economy.* Oxford: Basil Blackwell.
Dugan, J.T., T.S. McGrath, and R.B. Zelt. 1994. *Water-level changes in the High Plains Aquifer – Predevelopment to 1992.* US Geological Survey, Water Resources Investigations. Report 94-4027, Lincoln, Nebraska.
Duncan, M.L. 1987. High noon on the Ogallala Aquifer: Agriculture does not live by farmland preservation alone. *Washburn Law Journal* 27 (1): 16–103.
Emel, J., R. Roberts, and D. Sauri. 1992. Ideology, property, and groundwater resources: an exploration of relations. *Political Geography* 11 (1): 37–54.
Emel, J., and Gavin Bridge. 1995. The Earth as output. In Johnston, R., and M. Watts, eds, *The world in crisis*, 255–271. Oxford: Basil Blackwell.
Emel, J., and E. Brooks. 1988. Changes in the form and function of property rights under threatened resource scarcity. *Annals of the Association of American Geographers* 78: 241–42.
Emel, Jacque L., and Rebecca Roberts. 1995. Institutional form and its effect on environmental change: The case of groundwater in the Southern High Plains. *Annals of the Association of American Geographers* 85 (4): 664–683.
Farney, D. 1989. Abiding frontier: On the Great Plains, life becomes a fight for water and survival. *Wall Street Journal* 16 August, A5, p. 1.

Feder, Barnaby J. 1995. The standoff over beef: Some ranchers question meatpackers' big profits. *New York Times*, 17 October, D1, p. 10.
Fehrenbach, T.R. 1994. *Comanches: The destruction of a people.* New York: Da Capo Press.
Fenneman, N. 1931. *Physiography of the western United States.* New York: McGraw-Hill.
Fielding, William G. 1990. Meat industry competition and consolidation in the 1990s. Paper presented at the Annual Agricultural Outlook Conference, Outlook '91, Session #C17, US Department of Agriculture, Washington DC.
Findley, Rowe. 1991. Road to Santa Fe: The west's longest-lived trail still beckons to adventure. *National Geography Society*, March: 98–122.
Fitzsimmons, Margaret. 1989. Consequences of agricultural industrialization: Environmental and social change in the Salinas Valley, California, 1945–1978. Ph.D. dissertation, University of California at Los Angeles.
Fitzsimmons, M., J. Glaser, R. Monte Mor, S. Pincetl, and S. C. Rajan. 1994. Environmentalism and the liberal state. In O'Connor, 1994a: 198–216.
Forest, Bedford. 1996. Personal communication. Marketing Specialist, Texas Department of Agriculture, Amarillo.
Francis, J.G. 1990. Natural resources, contending theoretical perspectives and the problem of prescription. *Natural Resources Journal* 30: 263–282.
Foster, Morris W. 1991. *Being Comanche: A social history of an American Indian community.* Tucson: University of Arizona Press.
Foucault, Michel. 1972. *The archaeology of knowledge.* New York: Pantheon Books.
Friedman, A.E. 1971. Economics of the Common Pool: Property rights in exhaustible resources. *UCLA Law Review* 18: 858–887.
Friedmann, Harriet. 1982. The political economy of food: The rise and fall of the international food order. *American Journal of Sociology* 88: S248–S286.
Friedmann, Harriet, and Phillip McMichael. 1989. Agriculture and the state system: The rise and decline of national agricultures, 1870 to the present. *Sociologia Ruralis* 29 (2): 93–117.
Gale, R.P. and S.M. Cordray. 1994. Making sense of sustainability: Nine answers to what should be sustained. *Rural Sociology* 59 (2): 311–332.
Gard, W. 1959. *The great buffalo hunt: Its history and drama, and its role in the opening of the west.* Lincoln: University of Nebraska Press.
Genovese, Eugene. 1965. *The political economy of slavery.* New York: Pantheon.
Gibson, James S. 1932. Agriculture of the *Llano Estacado*. *Economic Geography* 8: 245–261.
Glover, Teresa, and Leland Southard. 1995. Cattle industry continues restructuring. *Agricultural Outlook*, December: 13–16. Washington DC: US Department of Agriculture.
Goldblatt, David. 1996. *Social theory and the environment.* Boulder: Westview Press.
Goldschmidt, W. 1978. *As you sow: Three studies in the social consequences of agribusiness.* Montclair, NJ: Allanheld, Osmun and Co.
Gordon, Joseph F. 1961. The history and development of irrigated cotton on the High Plains of Texas. Doctoral dissertation, Texas Technological College, Lubbock.
Gouveia, Lourdes. 1994. Global strategies and local linkages: The case of the US meatpacking industry. In Bonnano, Alessandro, Lawrence Busch, William

Friedland, Lourdes Gouveia, and Enzo Mingione, eds. *From Columbus to Con-Agra: The globalization of agriculture and food*. Lawrence: University Press of Kansas.

Grainger, Alan. 1990. *The threatening desert: Controlling desertification*. London: Earthscan.

Great Plains Drought Area Committee, 1936. Report of the Great Plains Drought Area Committee. August, Washington DC.

Green, Donald E. 1973. *Land of the underground rain: Irrigation on the Texas High Plains, 1910–1970*. Austin: University of Texas Press.

Green, Donald E. 1992. A history of irrigation technology used to exploit the Ogallala Aquifer. In Kromm and White, 1992a: 28–43.

Hacker, Caroline 1996. Personal communication, 26 June. Farm Service Agency, Hereford, Deaf Smith county, Texas.

Hagan, William T. 1976. *United States–Comanche relations: The reservation years*. New Haven and London: Yale University Press.

Hansen, J.M. 1991. *Gaining access: Congress and the farm lobby, 1919–1981*. Chicago: University of Chicago Press.

Hareven, Tamara K., and Randolph Langenbach. 1978. *Amoskeag: Life and work in an American factory city*. New York: Pantheon Books.

Haraway, Donna. 1997. *Modest witness @ second millennium. Femaleman meets OncoMouse*™. New York and London: Routledge.

Hassan, Zafar. 1993. Pakistan's textile industry is poised for downstream development. *Cotton International*: 181–184.

Hendee, J.C., and G.H. Stankey. 1973. Biocentricity in wilderness management. *Bioscience* 23: 535–538.

Higgins, K., and Barker, W. 1982. *Changes in vegetation structure in seeded nesting cover in the prairie pothole region. Special Scientific Report No. 242: Wildlife*. Washington, DC: US Department of Interior, Fish and Wildlife Service.

Hill, R. 1986. Economic, environmental and policy factors affecting cotton yields in the Texas High Plains. Master's thesis, Texas Tech University, Lubbock.

Hodge, Frederick W., ed. 1907, 1912. *Handbook of American Indians North of Mexico*. (2 vols.) Washington DC: Smithsonian Institution, Bureau of American Ethnology.

Holland, David, and Joe Carvalho. 1985. The changing mode of production in American agriculture: Emerging conflicts in agriculture's role in the reproduction of advanced capitalism. *Review of Radical Political Economics* 17 (4): 1–27.

HPUWCD #1 [High Plains Underground Water Conservation Distict]. 1983. Conservation through enhanced irrigation system performance. *The Cross Section* 29, no. 3: 2.

HPUWCD #1. 1995. Committees begin gathering information for regional water plan. *The Cross Section* 41 (1): 1.

HPUWCD #1. 1996a. Average decline of 1.91 recorded in district observation wells in 1995. *The Cross Section* 42 (4): 1.

HPUWCD #1. 1996b. Drought relief information meetings scheduled for West Texas cities. *The Cross Section* 42 (7): 1.

HPUWCD #1. 1996c. Legislators propose change to crop insurance rules as means of assisting producers stricken by drought. *The Cross Section* 42 (6): 1–2.

HPUWCD #1. 1997. Drought conditions may prompt water related legislation during session. *The Cross Section* 43 (1): 1.

HPUWCD #1. 1998. Texas Water Development Board publishes amended state water plan. *The Cross Section* 44 (5): 1–2, 4.

HPUWCD #1. 1999. 1999 Precipitation enhancement program ends. *The Cross Section* 45 (10): 1–2, 5.

Hyde, George. 1959. *Indians of the high plains: From the prehistoric period to the coming of the Europeans*. Norman: University of Oklahoma Press.

International Cotton Advisory Committee. 1988. *Cotton: Review of the world situation*. November–December 1987. Washington DC: ICAC.

Johnson, E. 1931. *The natural regions of Texas*. University of Texas Bulletin no. 1113. Austin: University of Texas Press.

Johnson, C.W. 1982. Texas Groundwater Law: A survey and some proposals. *Natural Resources Journal* 11: 1017–1029.

Johnson, Vance. 1947. *Heaven's tableland: The dust bowl story*. New York: Farrar, Straus and Co.

Johnson, W.D. 1899–1900. *The High Plains and their utilization. 21st annual report of the United States Geological Survey, Part IV (Hydrography)*: 599–741.

Johnson, W.D. 1900–1901. *The High Plains and their utilization. 22nd annual report of the United States Geological Survey, Part IV (Hydrography)*: 631–670.

Jones, Don L. 1954. Harvesting cotton with the mechanical stripper. *The Cotton Gin and Oil Mill Press* 3: 78–80.

Jones, Don L. 1959. Cotton on the Texas High Plains. *The Cotton Gin and Oil Mill Press* 10: 34–35.

Kanth, Rajani Kannepalli. 1997. *Breaking with enlightenment: The twilight of history and the rediscovery of utopia*. Atlantic Highlands, NJ: Humanities Press.

Kasperson, Jeanne X., Roger E. Kasperson, and B.L. Turner II, eds. 1995. *Regions at risk: Comparisons of threatened environments*. Tokyo: United Nations University Press.

Kasperson, R.E., J.X. Kasperson, B.L. Turner II, K. Down, and W.B. Meyer. 1995. Critical environmental regions: concepts, distinctions, and issues. In Kasperson, Kasperson, and Turner 1995: 1–41. Tokyo and New York: United Nations University Press.

Kavanagh, Thomas W. 1986. Political power and the political organization of Comanche politics, 1786–1875. Ph.D. dissertation, University of New Mexico.

Koontz, Stephen R., and Phillip Garcia. 1989. Meatpacker conduct in fed cattle pricing: An investigation of oligopsony power. *American Journal of Agricultural Economics* 75 (August): 537–548.

Kraenzel, C. 1955. *The Great Plains in transition*. Norman: University of Oklahoma Press.

Kromm, David, and Stephen White, eds. 1992a. *Groundwater exploitation in the High Plains*. Lawrence, Kansas: University Press of Kansas.

Kromm, David, and Stephen White. 1992b. The High Plains Ogallala region. In Kromm and White, 1992a: 1–27.

Kromm, David, and Stephen White. 1992c. Future prospects. In Kromm and White, 1992a: 224–9.

Lacewell, R.D., and G.S. Collins. 1984. Improving crop management. In Engelbert,

E.A., and A.F. Scheuring, eds, *Water scarcity: Impacts on western agriculture*, 180–203. Berkeley: University of California Press.

Lawson, M.P., and Stockton, C.W., 1981. Desert myth and climatic reality. *Annals of the Association of American Geographers* 71: 527–535.

Lee, J., B.L. Allen, R.E. Peterson, J.M. Gregory, and K.E. Moffett. 1994. Environmental controls on blowing dust direction at Lubbock, Texas, U.S.A. *Earth Surfaces Processes and Landforms* 19: 437–449.

Lind, R.C. 1982. A primer on the major issues relating to the discount rate for evaluating national energy options. In Lind, R.C., and K.J. Arrow, eds, *Discounting for time and risk in energy policy*. Baltimore: Johns Hopkins University Press.

Lines, M. 1995. Dynamics and uncertainty. In Folmer, H., H. Landis Gabel, and H. Opschoor, eds, *Principles of environmental and resource economics*. Aldershot: Edward Elgar.

Livingston, R.B. 1952. Relict true prairie communities in central Colorado. *Ecology* 33: 72–86.

Lubbock Chamber of Commerce. 1915. *An industrial survey of the city and county of Lubbock, Texas*. Lubbock: Lubbock Chamber of Commerce.

Lubbock Chamber of Commerce 1934. Application to the Texas Rural Homes Foundation for a farm community project of 100 units to be established on the South Plains of Texas in the vicinity of Lubbock, the county seat of Lubbock county, Texas. 15 February 1934.

Lubbock Chamber of Commerce. 1955. *Facts and figures about the city of Lubbock and the South Plains of Texas*. Lubbock: Lubbock Chamber of Commerce.

Lubbock Commercial Club. 1909. Short but forceful story of the plains and Lubbock County. Lubbock.

Maizels, Alfred. 1992. *Commodities in crisis: The commodity crisis of the 1980s and the political economy of international commodity policies*. Oxford: Clarendon Press.

Marsh, George Perkins. 1965 [1864]. *Man and nature, or, physical geography as modified by human action*, ed. David Lowenthal. Cambridge, MA: Harvard University Press.

Matthews, Anne. 1992. *Where the buffalo roam*. New York: Grove Weidenfield.

May, Kenneth. 1962. Plugging holes in irrigation is goal: Educating farmer to save water costly, takes time. *Lubbock Avalanche Journal*, Lubbock, 16 July 1962.

McMichael, Philip D. 1992. Tensions between national and international control of the world food order: Contours of a new food regime. *Sociological Perspectives* 35: 343–365.

Meadows, D.C. 1972. *The limits to growth*. London: Pan.

Mitchell, Robert, and Richard Carson, 1989. *Using surveys to value public goods: The contingent valuation method*. Washington DC: Resources for the Future.

Moench, Marcus. 1994. Approaches to groundwater management: To control or enable? *Economic and Political Weekly* 29, no. 30: 135–146.

Morris, David. 1991. *Cotton to 1996: Pressing a natural advantage*. Special Report No 2145. London: Economist Intelligence Unit.

Mortimore, Michael J. 1989. *Adapting to drought: Farmers' famine and desertification in West Africa*. Cambridge: Cambridge University Press.

Moseley, Lynn. 1996. Improved irrigation application efficiency just one benefit of center pivots. *The Cross Section*, 42, no. 11: 1–4.

Murphy, A. 1991. Regions as social constructs: The gap between theory and practice. *Progress in Human Geography* 15: 22–35.

Nall, Garry L. 1982. The cattle feeding industry on the Texas High Plains. In Dethloff and May, 1982: 10–115.

Nall, Garry L. 1990. Farming on the High Plains: Innovations and adaptation on arid lands in Texas. *Journal of the West* 29: 24–30.

NASS [National Agricultural Statistics Service]. 1996. *Texas cattle prices.* Washington DC: US Department of Agriculture.

Nordhaus, W.D. 1994. *Managing the global commons: The economics of climate change.* Cambridge, MA: The MIT Press.

Noyes, Stanley. 1993. *Los Comanches: the horse people, 1751–1845.* Albuquerque: University of New Mexico Press.

O'Connor, J. 1988. Capitalism, nature, socialism: A theoretical introduction *Capitalism, nature, socialism* 1 (1): 11–38.

O'Connor, Martin, ed. 1994a. *Is capitalism sustainable? Political economy and the politics of ecology.* New York: Guilford Press.

O'Connor, Martin. 1994b. Codependency and indeterminacy: critique of the theory of production. In O'Connor, 1994a: 53–75.

Odum, E.P. 1971. *Fundamentals of ecology.* 3rd edn. Philadelphia: W.B. Saunders.

Oliva, Leo E. 1989. The Santa Fe Trail in wartime: Expansion and preservation of the Union. *Journal of the West* 28: 53–58.

Opie, John. 1993. *Ogallala: Water for a dry land (a historical study in the possibilities for American sustainable agriculture).* Lincoln and London: University of Nebraska Press.

Ostrom, Eleanor. 1994. Self-organizing resource regimes: A brief report on a decade of policy analysis. *Common Property Resource Digest* 31: 14–19.

Pearce, D., E.B. Barbier, and A. Markandya. 1990. *Sustainable development: Economics and environment in the third world.* London: Earthscan.

Peckham, Darrell S., and John B. Ashworth. 1993. *The High Plains Aquifer System of Texas, 1980 to 1990: Overview and projections.* Austin: Texas Water Development Board.

Peet, Richard and Michael Watts. 1996. *Liberation ecologies: Environment, development and social movements.* London and New York: Routledge.

Pool, W.C. 1975. *Historic atlas of Texas.* Austin: Encino Press.

Popper, F., and D. Popper. 1991. The reinvention of the American frontier. *Amicus Journal* 13 (3): 4–7.

Popper, F., and D. Popper. 1994. Great Plains: Checkered past, hopeful future. *Forum for Applied Research and Public Policy* 9: 89–100.

Potter, D.M. 1954. *People of plenty: Economic abundance and the American character.* Chicago: University of Chicago Press.

Potts, F.P., J.L. Lewis, and W.L. Dorries. 1966. *Texas in maps.* Commerce: East Texas State University Research Publications.

Powell, J.W. 1879. *Report on the lands of the arid region of the United States.* Washington DC: Government Printing Office.

Powell, J.W. 1890. Institutions for the arid lands. *Century* 50: 111–116.

Pretty, J.N., and R. Howes. 1993. *Sustainable agriculture in Britain: Recent achievements and new policy challenges.* London: International Institute for Environment and Development.

Purcell, Wayne. 1990a. *Structural change in livestock: Causes, implications, alternatives*. Blacksburg, VA: Research Institute on Livestock Pricing.
Purcell, Wayne. 1990b. Economics of consolidation in the beef sector: Research challenges. *American Journal of Agricultural Economics* 72 (5): 1210–1218.
Radetzki, Maria. 1990. *A Guide to primary commodities in the world economy*. Oxford: Basil Blackwell.
Ragland, J.D. 1996. Personal communication, 26 June. Agricultural Extension Agent, Dimmit, Texas.
Rathjen, Frederick W. 1985. *The Texas Panhandle frontier*. Austin: University of Texas Press.
Redclift, M. 1988. Sustainable development and the market: A framework for analysis. *Futures*, December: 635–650.
Redclift, M. 1991. The multiple dimensions of sustainable development. *Geography* 76: 36–42.
Richardson, Rupert N. 1933. *The Comanche barrier to South Plains settlement: A century and a half of savage resistance of the advancing white frontier*. Glendale, CA: Arthur H. Clark Company.
Roberts, R., and J. Emel. 1992. Uneven development and the tragedy of the commons: Competing images for nature–society analysis. *Economic Geography* 68 (3): 249–271.
Robinson, Harriet. 1898. *Loom and Spindle: Or, life among the early mill girls*. New York: Thomas Y. Crowell.
Rollings, Willard H. 1989. *Indians of North America: the Comanche*. New York and Philadelphia: Chelsea House Publishers.
Sanderson, Steven E. 1986. The emergence of the "world steer": International and foreign domination in Latin American cattle production. In Sanderson, S.E., ed., *Food, the state, and international political economy: Dilemmas of developing countries*, 123–148. Lincoln: University of Nebraska Press.
Sanford, S.O. 1990. Trends in upland cotton exports. In *Proceedings of the Beltwide Cotton Production Research Conferences, January 9–14, 1990, Las Vegas, Nevada* (Memphis: National Cotton Council of America): 451–453.
Sayer, A. 1989. The "new" regional geography and problems of narrative. *Environment and Planning D: Society and Space* 7: 253–276.
Schell, O. 1985. *Modern meat: Antibiotics, hormones, and the pharmaceutical farm*. New York: Vintage Books.
Schlebecker, John T. 1963. *Cattle raising on the Plains 1900–1961*. Lincoln: University of Nebraska Press.
Secoy, William. 1953. *Changing military patterns of the Great Plains Indians (17th through 19th century)*. Lincoln and London: University of Nebraska Press.
Segarra, E., W. Keeling, and J.R. Abernathy. 1990. Analysis and evaluation of the impacts of cotton harvesting dates in the Southern High Plains of Texas. In *Proceedings of the Beltwide Cotton Production Research Conferences, January 9–14, 1990, Las Vegas, Nevada* (Memphis: National Cotton Council of America): 386–390.
Shaw, Lawrence. 1993. Roll with the Punches. *Cotton International*: 34.
Sheffy, Lester Fields. 1963. *The Francklyn Land and Cattle Company: A Panhandle enterprise*. Austin: University of Texas Press.

Shi, David E. 1985. *The simple life: plain living and high thinking in American culture*. New York: Oxford University Press.
Skaggs, Jimmy M. 1973. *The cattle trailing industry: Between supply and demand, 1866–1890*. Lawrence: The University Press of Kansas.
Smith, H.N. 1950. *Virgin lands*. Cambridge, MA: Harvard University Press.
Smith, Zachary. 1985. *Interest group interaction and groundwater policy formation in the Southwest*. Lanham, Maryland: University Press of America.
Smythe, W.E. 1911. *The conquest of arid America*. New York: Macmillan.
Southwestern Public Service Company. 1996. *1996 Fed cattle survey*. Amarillo: Southwestern Public Service Co.
Spencer, Bill. 1995. Hot Prices: How Long Will They Last? *Cotton Grower*, July: 29.
Spencer, William. 1995. "We Strongly Oppose Changes:" New ACSA President Ernst Schroeder says it's vital to our international trade to keep current farm legislation in place. *Cotton Grower*, May: 10–12.
Steer, A., and E. Lutz. 1993. Measuring environmentally sustainable development. *Finance and development*, December: 20–23.
Stephens, A. Ray. 1982. Trends in cattle raising in Texas since World War II. In Dethloff and May, 1982: 73–85.
Strange, J. 1984. The economic structure of sustainable agriculture. In Jackson, W., W. Berry, and B. Colman, eds, *Meeting the expectations of the land*. San Francisco: North Point Press.
Stroink, Nick. 1993. The Challenge. *Cotton International*: 31.
Successful Farming. 1995. Whatever happened to cattle cycles? September, B4, B10.
Takaki, Ronald. 1993. *Different mirror: A history of multicultural America*. Boston: Little Brown.
Taylor, P.W. 1986. *Respect for nature*. Princeton: Princeton University Press.
Templer, Otis W. 1985. Water conservation in a semi-arid agricultural region: The Texas High Plains. In Templer, Otis W., ed., *Forum of the Association for Arid Lands Studies*, vol. 1, 31–38. Lubbock: International Center for Arid and Semi-arid Land Studies, Texas Tech University.
Templer, Otis W. 1992. The legal context for groundwater use. In Kromm and White, 1992a: 64–87.
Texas Agricultural Experiment Station. 1968. *Agroclimatic atlas of Texas*. College Station: Texas A&M University.
Texas Department of Agriculture. 1910. Report of the Director on the establishment of the new state stations. *Bulletin #134*, 11/1910. College Station: Texas Agricultural Experiment Station.
TWDB [Texas Water Development Board] 1996a. *Meatpacking and feedlot water consumption data*. Austin: TWDB.
TWDB. 1996b. *Report 347: Surveys of irrigation in Texas, 1958, 1964, 1969, 1974, 1984, 1989, and 1994*. Austin: TWDB.
TWDB. 1997. *Water for Texas: A consensus based update to the state water plan*. Austin. TWDB.
Tharp, B.C. 1952. *Texas range grasses*. Austin: University of Texas Press.
The Irrigation Age. Various years, 1891–1902. Chicago: D.H. Anderson Publishing Co.
Thomas, Alfred B. 1940. *The Plains Indians and New Mexico, 1751–1778: A collec-*

tion of documents illustrative of the eastern frontier of New Mexico. Albuquerque: University of New Mexico Press.

Thrift, N. 1991. For a new regional geography 2. *Progress in Human Geography* 15: 456–465.

Tsai, Manfred. 1993. Looking to the PRC for Expansion. *Cotton International*: 43–50.

Ufkes, Frances M. 1993. Trade liberalization, agro-food politics and the globalization of agriculture. *Political Geography* 12 (3): 215–231.

Ufkes, Frances M. 1995. Lean and mean: US meat-packing in an era of agro-industrial restructuring. *Environment and Planning D: Society and Space*, 14: 683–705.

Underwood, Gary. 1991. Personal interview, May. Agent, Soil Conservation Service, Lubbock.

Urban, L.V. 1992. Texas High Plains. In Kromm and White, 1992a: 204–223.

Urry, J. 1992. The tourist gaze and the environment. *Theory, Culture and Society* 9: 1–26.

USDA [US Department of Agriculture]. 1941. *The Farm Security Administration*, 1 May 1941. Washington DC.

USDA. 1955. *Yearbook*. Washington DC.

USDA. 1988. *Census of Agriculture, Farm and ranch irrigation survey*. Washington DC: Bureau of the Census.

USDA. 1992–1994. *Census of Agriculture, Geographic Area Series, Texas*. Washington DC: Bureau of the Census.

USDA. 1996a. *Cattle and beef industry statistics*. Washington DC.

USDA. 1996b. *Cattle on feed*. Economic Research Service Animal Products. Washington DC.

USDA. 1996c. *Census of Agriculture, Farm and ranch irrigation survey*. Washington DC: Bureau of the Census.

USDA. 1996d. *Concentration in agriculture: A report of the USDA Advisory Committee on Agricultural Concentration*. June, 1996. Washington DC.

US Department of Commerce. 1925–1935. *Mid-decade Agriculture Census*. Washington DC: Bureau of the Census.

US Department of Commerce. 1900–1940. *Agriculture Census*. Washington DC: Bureau of the Census.

Verhovek, Sam Howe. 1996. Wheat farmers and ranchers are ruined. New York *Times* May 19, 1996: A1, B8.

Wallace, Ernest, and E. Adamson Hoebel. 1952. *The Comanches: Lords of the Southern Plains*. Norman: University of Oklahoma.

Walsh, J. 1980. What to do when the well runs dry. *Science* 210: 754–756.

Wang, Shumin. 1991. Expansion in North China: A total acreage of 6.7 million hectares and a lint yield of 6 million metric tons is the PRC's goal by the year 2000. *Cotton International*: 36–37.

Wang, Shumin. 1993. China's Cotton policy Fluctuates With Supply. *Cotton International*: 58.

Ward, Clement E. 1990. Structural change: Implications for competition and pricing in the feeder-packer subsector. In Purcell, 1990a: 98–119.

Watts, M. 1983. *Silent violence: Food, famine, and the peasantry in Northern Nigeria.* Berkeley: University of California Press.

Webb, W.P. 1931. *The Great Plains*. Boston: Ginn and Co.
West, Terry. 1990. USDA Forest Service management of the national grasslands. *Agricultural History* 64: 86–98.
Whatmore, S. 1993. Sustainable rural geographies? *Progress in Human Geography* 17 (4): 538–547.
White, R. 1991. *"It's your misfortune and none of my own": A new history of the American west.* Norman: University of Oklahoma Press.
White, Stephen E. 1994. Ogallala oases: Water use, population redistribution, and policy implications in the High Plains of western Kansas. *Annals of the Association of American Geographers* 84 (1): 29–45.
Widtsoe, J.A. 1911. *Dry-farming: A system of agriculture for countries under a low rainfall.* New York: Macmillan.
Worster, D. 1979. *Dust bowl: The southern plains in the 1930s*. Oxford: Oxford University Press.
Worster, Donald. 1985. *Nature's economy: A history of ecological ideas*. Cambridge and New York: Cambridge University Press.
Worster, D. 1994. *An unsettled country: Changing landscapes of the American west.* Albuquerque: University.
Wright, Gavin. 1978. *The political economy of the cotton south: Households, markets, and wealth in the nineteenth century.* New York and London: W.W. Norton.
Wright, N.G. 1988. Low-water-use native plants: Are they an economic alternative for commercial framing in the US Southwest? In Whitehead, E.E., C.F. Hutchinson, B.N. Timmermann, and R.G. Varady, eds, *Arid lands: Today and tomorrow*, 1341–1343. Boulder: Westview Press.
Young, Oran. 1982. *Natural resources and the state*. London: Unwin and Hyman.
Young, Thomas M. 1903. *The American cotton industry: A study of work and workers contributed to the Manchester Guardian*. New York: Charles Scribner's Sons.

Index